Engineering Risk and Hazard Assessment

Volume I

Editors

Abraham Kandel, Ph.D.
Chairman and Professor
Department of Computer Science
Florida State University
Tallahassee, Florida

Eitan Avni, Ph.D.
Research Scientist
Research and Development Division
Union Camp
Princeton, New Jersey

CRC Press
Taylor & Francis Group
Boca Raton London New York

CRC Press is an imprint of the
Taylor & Francis Group, an **informa** business

First published 1988 by CRC Press
Taylor & Francis Group
6000 Broken Sound Parkway NW, Suite 300
Boca Raton, FL 33487-2742

Reissued 2018 by CRC Press

© 1988 by CRC Press, Inc.
CRC Press is an imprint of Taylor & Francis Group, an Informa business

No claim to original U.S. Government works

Library of Congress Cataloging-in-Publication Data

Engineering risk and hazard assessment.

   Includes bibliographies and indexes.
   1. Technology--Risk assessment. I. Kandel, Abraham.
II. Avni, Eitan.
T174.5.E52   1988   363.1   87-20863
ISBN 0-8493-4655-X (set)
ISBN 0-8493-4656-8 (v. 1)
ISBN 0-8493-4657-6 (v. 1)

A Library of Congress record exists under LC control number: 87020863

ISBN 13: 978-1-315-89260-3 (hbk)
ISBN 13: 978-1-351-07170-3 (ebk)

Visit the Taylor & Francis Web site at http://www.taylorandfrancis.com and the
CRC Press Web site at http://www.crcpress.com

PREFACE

These volumes deal with the newly emerging field of "Risk and Hazard Assessment" and its application to science and engineering.

The past decade has seen rapid growth in this field but also some "real" disasters in both the U.S. and Soviet Union. It has been the recurrent nightmare of the 20th century: a nuclear power plant explodes, the core begins to melt, a conflagration ignites that spreads a radioactive cloud over the earth.

A malfunction of the coolant system in the core of the Soviet Union's Chernobyl reactor No. 4 may have triggered the violent chemical explosion in the Ukraine.

The Chernobyl disaster has inevitably renewed the debate over the safety of nuclear power plants far beyond Soviet borders. The worst U.S. disaster took place in 1979 at General Public Utility's Three Mile Island (TMI) plant near Harrisburg, Pa. All told, plants in 14 countries have recorded 151 "significant" incidents since 1971, according to a report by the General Accounting Office.

Scientists all over the world point out that there is a limit to how much safety technology can guarantee. Most accidents, including the one at TMI, involved a combination of equipment failure and human error. "In a population of 100 reactors operating over a period of 20 years, the crude cumulative probablility of a [severe] accident would be 45 percent," concluded a recent risk-assessment study by the Nuclear Regulatory Commission, which polices commerical reactors. John G. Kemeny, who headed President Carter's Commission on the TMI accident says: "Something unexpected can always happen. That's the lesson from TMI. All you can do is cut down on the probabilities." As many as half of the safety procedures now routinely conducted in the industry were born after 1979 in the wake of TMI. The industry itself has set up a self-policing body, the Institute of Nuclear Power Operations (INPO). For 15 years hazard analysis has been used in the chemical industry for comparing the risks to employees from various acute hazards. In this edited volume we try to take a better look at hazard assessment and risk analysis in order to improve our understanding of the subject matter and its applicability to science and engineering.

These volumes deal with issues such as short- and long-term hazards, setting priorities in safety, fault analysis for process plants, hazard identification and safety assessment of human-robot systems, plant fault diagnosis expert systems, knowledge based diagnostic systems, fault tree analysis, modeling of computer security systems for risk and reliability analysis, risk analysis of fatigue failure, fault evaluation of complex system, probabilistic risk analysis, and expert systems for fault detection.

It is our hope that this volume will provide the reader not only with valuable conceptual and technical information but also with a better view of the field, its problems, accomplishments, and future potentials.

Abraham Kandel
Eitan Avni
May 1986

THE EDITORS

Abraham Kandel is professor and Chairman of the Computer Science Department at Florida State University in Tallahassee, Florida. He is also the Director of The Institute for Expert Systems and Robotics at FSU. He received his Ph.D. in Electrical Engineering and Computer Science from the University of New Mexico, his M.S. in Electrical Engineering from the University of California, and his B.Sc. in Electrical Engineering from the Technion - Israel Institute of Technology. Dr. Kandel is a senior member of the Institute of Electrical and Electronics Engineering and a member of NAFIPS, the Pattern Recognition Society, and the Association for Computing Machinery, as well as an advisory editor to the international journals *Fuzzy Sets and Systems, Information Sciences,* and *Expert Systems.* He is also the co-author of *Fuzzy Switching and Automata: Theory and Applications* (1979), the author of *Fuzzy Techniques* in *Pattern Recognition* (1982), and co-author of *Discrete Mathematics for Computer Scientists* (1983), *Fuzzy Relational Databases — A Key to Expert Systems* (1984), *Approximate Reasoning in Expert Systems* (1985), and *Mathematical Techniques with Applications* (1986). He has written more than 150 research papers for numerous national and international professional publications in Computer Science.

Eitan Avni is a research scientist at Union Camp, Research and Development Division, Princeton, N.J. and was formerly an assistant professor of Chemical Engineering at Florida State University in Tallahassee, Fla. He received his Ph.D. in chemical engineering from the University of Connecticut. His current research interests include the application of fuzzy sets and artificial intelligence in chemical engineering, and risk and hazard assesment.

CONTRIBUTORS, VOLUME I

Eitan Avni, Ph.D.
Research Scientist
Union Camp
Princeton, New Jersey

L. T. Fan, Ph.D.
Professor and Head
Chemical Engineering Department
Kansas State University
Manhattan, Kansas

Ulrich Hauptmanns, Dr.-Ing.
Hauptprojektleiter
Projektbetreung
Gesellschaft für Reaktorsicherheit
Cologne, West Germany

Koichi Inoue, Dr. of Engineering
Professor
Department of Aeronautical Engineering
Kyoto University
Kyoto, Japan

Abraham Kandel, Ph.D.
Chairman and Professor
Department of Computer Science
Florida State University
Tallahassee, Florida

Trevor A. Kletz, D.Sc.
Professor
Department of Chemical Engineering
Loughborough University of Technology
Leicestershire, England

Hiromitsu Kumamoto, Dr. of Engineering
Research Associate
Department of Precision Mechanics
Kyoto University
Kyoto, Japan

F. S. Lai, Ph.D.
Research Leader
Engineering Section
Grain Marketing Research Laboratory
United States Department of Agriculture
Manhattan, Kansas

Yoshinobu Sato, Master of Engineering
Senior Researcher
Research Institute for Industrial Safety
Ministry of Labour
Tokyo, Japan

Sujeet Shenoi, M.S.
Graduate Research Assistant
Department of Chemical Engineering
Kansas State University
Manhattan, Kansas

CONTRIBUTORS, VOLUME II

Pedro Albrecht
Department of Civil Engineering
University of Maryland
College Park, Maryland

Eitan Avni
Research Consultant
Union Camp
Princeton, New Jersey

Wilker S. Bruce
Department of Computer Science
Florida State University
Tallahassee, Florida

Abraham Kandel, Ph.D.
Chairman and Professor
Department of Computer Science
Florida State University
Tallahassee, Florida

L. F. Pau, D.Sc.
Technical University of Denmark
Lyngby, Denmark

Ahmad Shafaghi, Ph.D.
Senior Engineer
Technica Inc.
Columbus, Ohio

W. E. Vesely, Ph.D.
Senior Staff Scientist
Science Applications International
Columbus, Ohio

Ronald R. Yager
Machine Intelligence Institute
Iona College
New Rochelle, New York

Nur Yazdani, Ph.D.
Assistant Professor
Department of Civil Engineering
College of Engineering
Florida A & M University
and
Florida State University
Tallahassee, Florida

Javier Yllera, Ph.D.
Consultant
Institute of Nuclear Engineering
Technical University
Berlin, West Germany

TABLE OF CONTENTS, VOLUME I

TABLE OF CONTENTS, VOLUME II

Chapter 1

NOW OR LATER?
A NUMERICAL COMPARISON OF SHORT- AND LONG-TERM HAZARDS

Trevor A. Kletz

TABLE OF CONTENTS

I. INTRODUCTION

In the Loss Prevention Symposium held in Newcastle in 1971, I[1] presented one of the first papers on the use of numerical methods for comparing the different risks to which employees in the process industries are exposed. Since then such methods have been used for setting priorities between different acute risks and the literature on the subject is now extensive.[2-10a] In this chapter, I suggest an extension of these methods to hazards which take a long time, typically several decades, to produce their effects.

Consider a substance X, the product of an industrial process, which can cause harm in two distinct ways:

1. It may leak out of the plant, vaporize, mix with air, and be ignited, thus injuring or killing people by fire or explosion. Very small leaks do not matter: the hazard arises only if the leak exceeds several kilograms and is unlikely to be serious unless it exceeds 1 tonne (though smaller leads can be serious in confined spaces).

2. Exposure of employees to small quantities of the vapor for long periods for many years may cause industrial disease which may lead to premature death. In general, we are not concerned with the occasional large release but with the small quantities present in the atmosphere as the result of minute leaks from joints, glands, sample and drain points, maintenance operations, and so on. (However, for some materials occasional large doses may produce long-term effects, or may cause sensitization). The Threshold Limit Value (TLV) gives the time-weighted average (TWA) concentration for a normal 8-hr workday or 40-hr work week, to which nearly all workers may be repeatedly exposed daily, without adverse effect. It does not, of course, provide a sharp division between safe and unsafe conditions. In the U.K., TLVs are now being replaced by control or recommended limits. The concentration should be decreased below these when it is reasonably practicable to do so. (Control limits have greater force and are set when there is sufficient evidence to justify them; recommended limits [usually the old TLVs] are set for more substances.)

We try, of course, to prevent both sorts of leaks and we spend a great deal of money and effort in doing so. How does our success in overcoming hazard (1) compare with our success in overcoming hazard (2)? We do not know. Should we put more effort into preventing the occasional big leaks or more effort into preventing the continuous small leaks? We do not know. Usually different people, using different criteria and financed by different budgets are responsible for dealing with the two hazards. Finding a way of talking to them both at the same time is like finding a way of communicating, at the same time, to people who speak different languages. This is attempted in this chapter.

The problem is not just one of comparing risks. It is made more difficult by the fact that the risks are different and are felt to be different by those who are subjected to them. Suppose that the probabilities of an employee being killed by the acute hazard (1) and the chronic or long-term hazard (2) are equal. An employee might feel that hazard (2) leaves him with an extra 20 or more years of life and, therefore, further resources should be spent on reducing the risk from hazard (1). On the other hand, another employee might feel that a fire or explosion is soon over while industrial disease may mean years of worry, wondering whether or not he will contract it, possibly followed by many years of illness, and reduced quality of life.

From society's point of view, long-term hazards, though they may kill many people, will kill them over a long period of time and will not produce the same trauma or public outcry as one that kills many people at a time.

In one important respect, the problem of long-term effects differs from all other industrial problems — often the size of the problem is not known. As we shall see in Section VI, we do not know how many people die or suffer from industrial disease (of all sorts, not just the legally prescribed diseases), because the same diseases usually also have nonoccupational causes. Before we spend resources on, for instance, reducing the usage of raw materials or energy, improving product quality, or preventing accidents, we start by asking ourselves what is the present usage, quality, accident rate, etc., and what is the scope for improvement. When dealing with toxicological hazards we do not seem able to do this.

A Personal Note

I have some experience of industrial accidents, but little knowledge of toxicology. In trying to compare the two I may be accused, like everyone who tries to compare two subjects normally considered apart, of dabbling in a field in which I am no expert. I admit the accusation, but if we want to knock a hole in a wall, we have to start from one side. Some may argue that immediate deaths and delayed deaths are so different that they cannot be compared; however, comparing different things is what management is about. Managers have to set priorities between different sorts of tasks. The question is not whether we compare short- and long-term hazards but whether we do so openly and explicitly or inwardly, using unknown criteria (unknown to ourselves as well as others).

Let us look at some examples, starting with those where the size of the problem is known.

II. COAL DUST

This substance is not of great interest to the process industries, but it is the cause of serious industrial disease; in 1975, of 802 deaths from prescribed industrial disease in the U.K., 643 were due to pneumoconiosis, most of it caused by coal dust, and it is one of the few substances for which the dose-response relationship is known. The National Coal Board has shown[11] that

$$P = \sin^2 (0.0704x - 0.1201)$$

where x = mean respirable dust concentration in milligrams per cubic meter, and P = probability of developing category 2/1 pneumoconiosis (ILO classification) in 35 years. When calculating the sine, the expression in the brackets is assumed to be in radians.

This equation is based on a statistical extrapolation of observations over 10 years in 20 coal mines and mean coalface dust concentrations in those mines up to 8 mg/m^3.

In 1970, the National Coal Board set standards which implied that long-term concentrations experienced by individuals would not exceed 4.3 mg/m^3. The corresponding value of P is 9.42×10^{-4} per person per year. This would be equivalent to a fatal accident rate (FAR) of 50 or 100 deaths per 10^5 men per year, if we assume that all cases of category 2/1 pneumoconiosis lead to premature death.

However, by no means do all cases of category 2/1 pneumoconiosis lead to premature death. According to the Institute of Occupational Medicine,[12] over 22 years, men in the 25 to 34 year age group with categories 1, 2, and 3 pneumoconiosis had survival rates of 90.1% compared with 93.0% for those with category 0. The difference, they state, is statistically significant and suggests that the mortality risk for those with simple pneumoconiosis is about 60% higher than that experienced by young men with no radiological signs initially.

What happens after 22 years is not known. Let us therefore assume that ultimately, one third of those with simple pneumoconiosis dies prematurely, though the time of death may be many years ahead. The pneumoconiosis death rate is then equivalent to a FAR of 17 or 33 deaths per 10^5 men per year. For comparison, the FAR for all acute accidents in coal mines was 40 (80 deaths per 10^5 men per year) in the 1960s but fell to about 14 (28 deaths per 10^5 men per year) in the 1970s. A coal miner joining the industry is thus about as likely to die from pneumoconiosis as from an acute accident, if past trends continue. This suggests that the Coal Board has allocated the priorities between short- and long-term hazards almost correctly.

However, for a man age 20 to 24, the total probability of death from all causes (including natural causes) is 10^{-3}/year.[13] If he is a coal miner, acute industrial accidents increase this by a factor of 3. Pneumoconiosis also increases his risk of death by a factor of 3, but this will probably not take effect for about 40 years, by which time his total probability of death from all causes will have risen by 20 times to 20×10^{-3}/year. Viewed in this way, the risk of death from pneumoconiosis does not seem as bad as the risk of death from acute industrial accidents.

On the other hand, it could be argued that our hypothetical coal miner may have looked forward to his retirement for 40 years and that death at the time of his retirement is worse than an early death.

This example shows that we cannot assume that the probability of death is necessarily the right parameter for use in comparing immediate and delayed effects. Many people may consider that delayed death is preferable; others may take the opposite view. The actions of the National Coal Board imply that an immediate death is equivalent to a delayed death. Whatever our views, they seem to have the balance between acute and chronic risks well within an order of magnitude. (Incidentally, halving the coal dust concentration will decrease the incidence of 2/1 pneumoconiosis by about 20 times).

III. RADIATION

This is another area where good dose-response data are available, though there is some doubt as to whether it can be extrapolated to low concentrations. For employees, the agreed maximum dose is 50 (millisieverts) mSv/year (5 rem/year) which is believed to give a risk of death of 5×10^{-4}/year (FAR 25 or 50 deaths per 10^5 men per year). This seems high compared with acute risks, particularly as people exposed to radioactivity are presumably also exposed to normal acute industrial risks as well. If we assume that the FAR for these acute risks is 2 (as in the chemical industry and most U.K. manufacturing industries), then those who set nuclear standards are implying that an immediate death is about 12 times worse than a delayed death. Kinchin,[14,15] a former head of the Safety and Reliability Directorate of the U.K. Atomic Energy Authority, has suggested levels of risk which are so low that action to reduce them further is not justified; his figure for delayed deaths is 30 times higher than his figure for immediate deaths. Of course, the average exposure is well below the maximum permitted but if one is exposed to the maximum this is not much consolation.

Reissland and Harries[16] have suggested, taking radiation as an example, that loss of expectation of life should be used for comparing risks instead of the probability of death. Table 1, based on their data, shows that for a man age 35 years or more and exposed to 50 mSv/year, the acute risks are worse than the chronic risks, assuming as before that he is exposed to the same acute risks as in other industries. However, as Griffiths[17] points out, loss of life expectancy is misleading. Consider a group of 1000 men. All of them could lose 10 days of life or one man could lose 10,000 days (27 years), but the average loss of life expectancy is the same. In no way can the two cases be considered similar. For a man who is unemployed, the unemployment rate is 100%.*

---

*   Since this chapter was written, estimates of the probability that radiation will cause death have increased. The International Commission on Radiological Protection (ICRP) now estimates that the probability is 25% higher than noted in this chapter and a recent report states: "It is prudent to assume that the results of a review of the doses received by the Japanese atomic bomb victims could increase the estimated risk of fatal cancer by a factor of up to two." (Layfield, F., *Sizewell B Public Inquiry: Report*, Vol. 1, Her Majesty's Stationery Office, London, 1987, paragraph 2.70(c).

Table 1
LOSS OF EXPECTATION OF LIFE FOR VARIOUS RISKS

Risk[a]	FAR	Deaths/ 10⁵men/ year	Loss of expectation of life (days); age at beginning of exposure (years)				
			20	**30**	**40**	**50**	**60**
Acute: U.K. manufacturing industry[b]	2	4	20	13.5	8	4	1
Chemical industry	4	8	41	27	16	8	2
Chronic exposure to 50 mSV radiation/year	25	50	68	32	12	3	0.5

[a]   Assuming exposure continues for rest of working life.
[b]   All premises covered by U.K. Factories Act.

IV. ASBESTOS

From about 1969 until 1985, the U.K. TLV for white asbestos (chrysotile) was two fibers per milliliter. This was set following a study by the British Occupational Hygiene Society (BOHS)[18] which showed that 1% of those exposed to this concentration for a working lifetime would develop asbestosis. Later work by BOHS[18a] showed that the risk is greater. According to Peto,[18b] 5 to 10% of men exposed to one to two fibers per milliliter for 50 years will die prematurely; the maximum allowable concentration in the working atmosphere has now (since 1985) been reduced in the U.K to 0.5 fibers per milliliter. (At the same time, the TLV has been replaced by a control limit. While TLVs were considered acceptable, and there was no obligation to go below these limits, the concentration in the workplace atmosphere must be reduced below the control limit if it is "reasonably practicable" to do so).

According to Peto[18b] the old TLV corresponded to an FAR of 50 to 100 (though this was not realized at the time it was set). The new control limit (assuming a linear response) will correspond to an FAR of 25 (50 deaths per 10⁵ men per year). This seems high compared with the acute risk to which asbestos workers are also exposed, but actual doses even for the most highly exposed individuals may be a good deal lower than the present-day control limits (although they were not in the past). Unfortunately, very little data are available on the ratio between actual exposures and maximum permitted exposures, but Doll and Peto[18c] estimate a ratio of 1/2 for asbestos. (The same figure has been estimated for coal dust.[11]) Even assuming this ratio, the FAR is such that efforts to move below the control limit are well justified. The position is complicated by the fact that smokers are much more at risk than nonsmokers. The simplest way of reducing the FAR would be to employ only non-smokers where asbestos is handled.

V. CHEMICALS

For chemicals, very little accurate dose-response data are available. Roach[19] pointed out that there are 612 substances in the 1976 list of TLVs published by the American Conference of Government Industrial Hygienists. Of these, 63 TLVs are based on industrial surveys, of which only eight involve more than 200 people, and 50 are based on laboratory studies on volunteers.

This sparsity of data is often quoted as a reason for not using hazard analysis;[20] however, TLVs are set despite the lack of data, and if there is enough information to set a TLV it should be possible to estimate a probability of death. The probability may err on the safe side, but even so, we can still compare this probability with that of acute accidents. (Some TLVs are based on the concentration required to cause irritation, not risk to life. These are not the ones with which I am concerned. When a TLV is based on a risk to life, if it is possible to set a TLV, it should be possible to estimate the risk to life).

In applying hazard analysis to acute problems it is also often necessary to use expert judgment to fill gaps in the data. If we break problems down into component questions (What was the dose of A? For how long did it continue? What is the effect of such a dose?), answering them with facts where possible and with expert opinion when no facts are available, we are more likely to get a correct answer than if we try to answer the whole problem with expert opinion. Such opinions should be used as substitutes for unavailable data rather than as an alternative method of problem solving.

An attempt to compare acute and long-term risk has been made for workers exposed to benzene.[21] It shows that exposure to 10 to 1000 ppm produces FARs of 1 to 10 (2 to 20 deaths per 10^5 men per year), but that there is no evidence that exposure below 10 ppm produces any excess risk. This is confirmed by a recent extensive review of the subject by Fielder.[22]

This brings us to the question of linearity on which expert opinion is divided.[23] Some writers believe that the effect of a toxic chemical is proportional to the dose and that even one molecule (or one photon of radiation) can cause disease, though the probability is small.[24] Others consider that there is a threshold dose below which there is no effect, the body's natural defense mechanism coping with the alien material; only when the body's defense mechanism is overwhelmed does disease result. It may be that the response is linear for some substances or effects (e.g., carcinogens), but that there is a threshold for others (e.g., coal dust).

The argument cannot be settled epidemiologically, because at low doses the effects are too small to be detectable above the background noise, i.e., the naturally occurring incidence of the disease.

In most cases a TLV may be exceeded for only a short time. A man should not be exposed to twice the TLV for 4 hr and then to a 0 concentration for another 4 hr. This implies that the dose-response data are not or may not be linear at these concentrations, and that low doses produce a lesser effect than would be expected by extrapolation from high doses. However, in making comparisons of the type discussed in this chapter, it would be prudent to base them on a linear hypothesis.

VI. ALL INDUSTRY

Can we derive any useful information by looking at U.K industry as a whole? From 1975 to 1978 the average number of deaths from industrial accidents was 694 per year and the average number of deaths from prescribed industrial disease was 925 per year (1974 to 1978).[25] This suggests that across industry as a whole, accidents and disease are problems of comparable magnitude; however, about 800 deaths per year out of the 925 are due to pneumoconiosis, asbestosis, and byssinosis, diseases restricted to a few industries and occupations. This suggests that acute accidents are the major problem.

The figures for industrial disease include only those due to prescribed disease, however.[26] Some deaths will be due to nonprescribed diseases, but there is no consensus of opinion on the number. Considering cancer alone, one report[27] estimated that 1% of male deaths from cancer (i.e., 0.3% of all male deaths) or 1000 per year might have occupational causes. Other reports[28-30] have suggested much larger figures. Thus the NCI/NIEHS/NIOSH report[29] states that 5% of all male deaths in the U.S. are due to occupational cancers, while the U.K. ASTMS report[30] suggests 3 to 6% (10,000 to 20,000 per year) for the U.K. These latter reports have been extensively critisized.[31] ("It shows how a group of reasonable men can collectively generate an unreasonable report.")

The Royal Society report[27] seems more likely to be correct. The 1000 male deaths from occupational cancer per year will include some of those due to prescribed disease, and there will be some deaths due to nonprescribed diseases besides cancer; hence, it seems that the total number of deaths per year due to occupational disease is perhaps between 1000 and 2000. The uncertainty in the data is illustrated by the fact that the Trades Union Congress states, "Every year about 1400 people die as a result of occupational accidents and diseases,"[32] while Farmer writes, "on a conservative estimate (cancer) kills twice as many workers as industrial accidents".[33] Death from occupational disease is thus a problem comparable with, and perhaps rather worse than, death from industrial accidents and justifies appropriate allocation of resources. Note that deaths from industrial disease reflect the working conditions of many years ago.

If we compare days lost due to industrial accidents with days lost due to industrial disease, we find that in 1976—1977 in the U.K accidents caused 12.2×10^6 lost days, while prescribed diseases caused 0.5×10^6 lost days.[34] Even if we multiply by ten to allow for nonprescribed diseases, industrial disease, measured in this way, seems a smaller problem than industrial accidents.

VII. CONCLUSIONS

This chapter has explored, in a preliminary way, the possibility of using estimates of the probability of death for comparing acute and chronic hazards. The purpose of doing so is to help decide whether reduction of acute risks or reduction of chronic risks should have priority. Results suggest that the allocation of resources between the two types of hazard is probably not too far-fetched.

A major difficulty is knowing whether delayed death is worse than immediate death. On one hand, we have 20 to 40 more years of life; on the other hand, we have the worry of possible disease and years of illness before death. Perhaps these two effects can be offset and all deaths treated as equally undesirable?

We do need a new look at the data on which TLVs are based in order to estimate (however roughly and with a safety factor to cover uncertainties) the probability of death at various exposures. Until we can do this we do not know if we are setting TLVs too low and thus spending on industrial hygiene resources that would be better spent on the reduction of acute hazards, or vice versa.

The evidence suggests, however, that in many industries the acute and chronic risks are now within an order of magnitude of each other. This may not seem very good, but it is not bad for problems of resource allocation. Nationally, for example, our allocation of resources to medical care, road safety, industrial safety, and discouragement of smoking bears no relation at all to either the relative risks or the costs of reducing the hazards.[2,2a]

VIII. FURTHER WORK

In this chapter the relative sizes of short- and long-term risks have been compared and it has been assumed that if we can identify the higher risks, we should give priority to their reduction. There is, however, another way of comparing risks. We can give priority to the expenditure which will save most lives per $1 million spent. It has been argued elsewhere[2,2a] that this method should not be preferred as it leads to the toleration of risks which are high but expensive to reduce, but that it may be useful as a secondary criterion. As to the present case, there is no information on the relative costs of reducing short- and long-term risks. This is a subject worthy of further investigation.

ACKNOWLEDGMENTS

Thanks are due to the many colleagues who suggested ideas for this chapter or commented on the draft. The opinions, however, are the author's.

Thanks are also due to the U.K. Science and Engineering Research Council for financial support.

This paper was originally presented at the 4th International Symposium on Loss Prevention and Safety Promotion in the Process Industries, September 1983, but the section on asbestos has been revised. It is reproduced by permission of the Institution of Chemical Engineers, U.K.

REFERENCES

1. **Kletz, T. A.**, *Int. Chem. Eng. Symp. Ser.*, 34, 75, 1971.
2. **Kletz, T. A.**, *Chemical Engineering in a Changing World*, Koetsier, W. T., Ed., Elsevier, Amsterdam, 1976, 397.
2a. **Kletz, T. A.**, Setting priorities in safety, in *Engineering Risk and Hazard Assessment*, Kandel, A. and Avni, E., Eds., CRC Press, Boca Raton, Fla., 1988.
3. **Kletz, T. A.**, *Hydrocarbon Process.*, 56(5), 297, 1977.
4. **Kletz, T. A.**, *Chem. Eng. Prog.*, 72(11), 48, 1976.
5. **Kletz, T. A.**, *Reliability Eng.*, 1(3), 35, 1981.
6. **Lees, F. P.**, *Loss Prevention in the Process Industries*, Butterworths, London, 1980, chap.9.
7. **Lawley, H. G.**, *Chem. Eng. Prog.*, 70(4), 45, 1974.
8. **Lawley, H. G.**, *Reliability Eng.*, 1(2), 89, 1980.
9. **Gibson, S. B.**, *Chem. Eng. Prog.*, 72(2), 59, 1976.
10. **Stewart, R. M.**, *Int. Chem. Eng. Symp. Ser.*, 34, 99, 1971.
10a. **Kletz, T. A.**, *Hazop and Hazan — Notes on the Identification & Assessment of Hazards*, Institution of Chemical Engineers, Rugby, U.K., 1983.
11. **Jacobsen, M., Rae, S., Walton, W. H., and Regan, S. M.**, *Inhaled Particles*, Vol. 3, Walton, W. H., Ed., Unwin Bros., 903.
12. Private Communication, A report is being prepared for publication.
13. **Lawrence, E., Ed.**, *Annual Abstract of Statistics*, Her Majesty's Stationery Office, London, 1981, Table 2.32.
14. **Griffiths, R. F.**, *Atom*, p. 3, December 1970.
15. **Kinchin, G. H.**, *Proc. Inst. Civ. Eng.*, Vol. 64, (Part 1), 431, 1978.
16. **Reissland, R. and Harries, V.**, *New Sci.*, p. 809, September 13, 1979.
17. **Griffiths, R. F.**, *Dealing with Risk*, University of Manchester Press, Manchester, England, 1981, chap. 4.
18. Committee on Hygiene Standards, British Occupational Hygiene Society, *Ann. Occup. Hyg.*, 11, 47, 1968.
18a. Committee on Hygiene Standards, British Occupational Hygiene Society, *Ann. Occup. Hyg.*, 27, 55, 1983.
18b. **Peto, J.**, *Lancet*, 1, 484, 1978.
18c. **Doll, R. and Peto, J.**, *Asbestos: Effects on Health of Exposure to Asbestos*, Her Majesty's Stationery Office, London, 1985.
19. **Roach, S. A.**, Control limits (workplace environment), National Health and Safety Conference, 1980, Victor Green Publ., London.
20. **Wood, C., Ed.**, *Human Health and Environmental Toxicants*, Int. Cong. Symp. Ser. No. 17, Royal Society of Medicine, London, 1980.
21. **Mountfield, B. A.**, A Behavioural Approach to the Assessment of Occupational Risk, M.Sc. dissertation, London School of Hygiene and Tropical Medicine, 1978.
22. **Fielder, R. J.**, *Toxicity Review 4: Benzene*, Her Majesty's Stationery Office, London, 1982.
23. **Truhaut, R.**, *Am. Ind. Hyg. Assoc. J.*, 41(10), 685, 1980.

24. **Baldwin, R. H.,** *Chem. Tech.*, 9(3), 156, 1979.
25. Health and Safety Executive, Health and Safety Statistics 1978—1979, Her Majesty's Stationery Office, London, 1981, 12 and 63.
26. Health and Safety Executive, Health and Safety Statistics 1978—1979, Her Majesty's Stationery Office, London, 1981, 64.
27. *Long-Term Toxic Effects: A Study Group Report,* The Royal Society, London, 1978.
28. **Epstein, S.,** *The Politics of Cancer,* rev. ed., Anchor Press, 1979.
29. National Cancer Institute/National Institute of Environmental Health Sciences/National Institute for Occupational Safety and Health, *Estimates of the Fraction of Cancer in the U.S. related to Occupational Factors,* NCI, Frederick, Md., 1978.
30. ASTMS, *The Prevention of Occupational Cancer: Policy Document,* 1980.
31. **Peto, R.,** *Nature (London),* 284, 297, 1980.
32. *Workplace Health and Safety Services,* U.K. Trades Union Congress Publ., London, 1980.
33. **Farmer, D.,** *Health and Safety at Work,* 9, 18, 1982.
34. Social Security Statistics, U.K. Department of Health and Social Security, London, 1977, Table 20.70.

Chapter 2

SETTING PRIORITIES IN SAFETY

Trevor A. Kletz

TABLE OF CONTENTS

I. INTRODUCTION

It is now widely accepted that not everything possible can be done to prevent every conceivable risk to life, however unlikely or trivial. Sufficient resources are never available. The decisions to be made are these: which risks to remove or reduce first, and which to leave until later, perhaps indefinitely. In other words, we have to set priorities.

There are various ways of setting these priorities. When the public and politicians are involved, it often seems that those who shout the loudest get the most. If a more systematic and more defensible method is found, then there are two distinct approaches: remove or reduce first those risks which exceed a particular level or remove or reduce first those risks that are cheapest to remove or reduce. This chapter describes and compares the two methods with particular reference to the process industries, though much of what will be said is relevant to other industries.

II. TARGET SETTING

This is the method most widely used in industry, whether one is concerned with risks to employees or with risks to the public. The principle can be illustrated by considering handrails around places of work from which someone may fall 6 ft or more. According to the U.K. Factories Act these have to be between 3 ft and 3 ft 9 in. high. A sort of intuitive hazard analysis has shown that at this height the chance of falling over them, though not zero, is so small compared with all the other risks to which people are exposed, that it is hardly worth worrying about. One can hardly imagine a situation in which lower handrails were accepted on unprofitable plants, and higher ones demanded on very profitable plants. Similarly, the thickness of shielding around radioactive materials is adjusted so that no one receives more than an agreed amount of radioactivity.

Most of the criteria or targets that are set for risk are more complex but the principle is the same. In industry investigators often try to estimate the risks to life to which employees are exposed and then try to remove or reduce those risks which exceed an average or specified value.[1-3] When the public is involved, the average risk to life of those who live near the factory is estimated, or the risk to the person at greatest risk, or the probability that 10, 100, 1000, or more people may be killed, and these are compared with a target or criterion. If the risks are above the target, action must be taken to reduce them. If the risks are below the target, then there is no reason the factory should not be extended. The various targets or criteria that have been proposed are discussed elsewhere.[4]

Ideally, these targets should be set so that the levels of risk are comparable; within the factory money should not be spent on raising the height of the handrails if the risk of falling over them is smaller than the risk of being harmed by a fire or a toxic gas leak. In the wider field, no one should spend resources on reducing risk to the public from plant A if the risk from plant B is similar in nature but greater. The largest risks should be dealt with first. In practice, of course, the targets often differ by orders of magnitude, the nuclear industry, for example, aiming for lower risks than, say, the oil or chemical industries, which in turn aim for much lower risks than long-established industries such as agriculture or coal mining.

A variation on the target setting approach is to have two targets or criteria. An upper level of risk which should never be exceeded — if the risk cannot be reduced below this level, then the activity which produces the risk must be stopped — and a lower level which investigators try to reach but are not compelled to reach; all must reach it if, to use the legal phrase, it is "reasonably practicable" to do so. If this lower level is reached, then it is not worthwhile expending any effort on reducing the risk further.[5]

This two-level approach can be illustrated by considering concentrations of toxic or harmful gases or dusts in the workplace atmosphere. At one time, the published threshold limit values (TLVs) were considered to be concentrations which could safely be breathed for 8 hr/day, year in, year out.[6]

Though they never had the full force of law, failure to achieve them could be considered as evidence of failure to provide a safe place of work. If the TLVs could be reached, there was no need to try to get below them.

The approach has now changed. The TLVs have been replaced by control or recommended limits which should not be exceeded.[7] (Control limits have greater force and are set when there is sufficient evidence to justify them; recommended limits, usually the old TLVs, are set for more substances). However, there is still an obligation to go below them if it is "reasonably practicable" to do so. For asbestos a "lower action level" has been proposed[8] below which no action is required and the greater use of such lower levels has been advocated.[9]

Note that a company has been prosecuted and convicted for not doing all that was "reasonably practicable" to remove asbestos from the working atmosphere, although no evidence was submitted that the concentration exceeded the control level or any other level.[10]

III. "ACCEPTABLE RISKS"

The risk targets or criteria are sometimes called acceptable risk levels, but I do not like this phrase. No one has the right to decide what risks are acceptable to other people and no person should ever knowingly fail to act when others lives are at risk, but everything cannot be done at once; priorities have to be set.

More pragmatically, particularly when talking to a wider audience than fellow technologists, the use of the phrase "acceptable risk" often causes people to stop listening. "What right have you," they say, "to decide what risks are acceptable to me?" Everyone has problems with priorities; most people realize that we cannot do everything at once, and they are more likely to listen if we talk about priorities. We do not, of course, remove priority problems by asking for more resources. We merely move the target level to a different point.

IV. REMOVING FIRST THE RISKS THAT ARE CHEAPEST TO REMOVE

In this alternative approach the largest risks are not removed first, nor those that exceed a target level, but those that are cheapest to remove. Therefore, more lives are saved per dollar spent. As this approach is more cost-effective, why is it not normally used in industry? There are several reasons; the first is moral. An employee or a member of the public may accept that a risk is so small, compared with the other risks around, that it is hardly worth worrying about, but he or she will hardly accept a risk because it is expensive to remove. It may be better for society as a whole, but not for him or her.

Restating the same objection in other words, although the total number of people killed in an organization or society may be reduced by concentrating the risks on a few individuals, no one is prepared to do so. The preference is to spread the risks more or less equally, or at least ensure that no one is exposed to a level of risk that would be regarded as significant.

Note that in industry the lives saved are notional. If money is spent on reducing a particular risk, the only result is making the already low risk of an accident even lower. It is unlikely that anyone's life will actually be saved and this makes it easier to adopt the moral attitude just described. In road safety, on the other hand, we are dealing with real lives; more lives will actually be saved if money is spent in a more cost-effective way, and in this field of activity attempts are made to spend money in ways that do save the most lives per dollar spent. One does not try to equalize the risks between different categories of road user, though it might be argued that pedestrians, who are exposed against their will, should be subjected to a lower risk.

The second reason is pragmatic. If it is agreed to remove risks that are inexpensive to remove, but to accept those that are expensive to remove, then there is a temptation for every design engineer and manager to say that the risks on his plant are expensive to remove. If, however, all risks must be reduced below a certain level, then experience shows that design engineers and plant managers do find "reasonably practicable" ways of reducing them below that level.

A third reason is that as already stated, the usual procedure in industry has always been to work to a risk, not a cost, criterion.

Despite these comments, the cost of saving a life is useful in industry as a secondary criterion. If the notional cost of saving a life is greatly in excess of the normal for the industry, the usual criterion should not be exceeded, but a cheaper solution should be sought. Experience shows that in practice it can usually be found. There is usually more than one solution to every problem.

V. WEIGHING IN THE BALANCE

A variation on removing the risks that are cheapest to remove is weighing in the balance or trading-off. Compare the costs of various courses of action against the benefits, expressing them all in common units (usually money), and adopt the course of action that produces the largest net benefit (or smallest net detriment). For example, compare the cost per year of preventing an accident with the cost of the damage it will produce, multiplied by the probability per year that it will occur, or compare the cost of preventing pollution with the value of the damage caused by the pollution. This approach is widely used, data permitting, when death is not involved. However, when fatal accidents are considered there is a reluctance to use it because there is no generally accepted value of a life, though many attempts have been made to estimate values (see Section VI). An advantage of this approach is that it causes one to identify and value the benefits to be derived from human activity instead of our taking them for granted.

In the U.K., weighing in the balance is, in theory, sanctified by the law. The words "reasonably practicable" which occur in the Factories Act (1961), the Health and Safety at Work Act (1974), and earlier legislation have been defined as implying "that a computation must be made in which the quantum of risk is placed on one scale and the sacrifice involved in the measures necessary for averting the risk (whether in money, time or trouble) is placed in the other, and that, if it be shown that there is a gross disproportion between them — the risk being insignificant in relation to the sacrifice — the defendants discharge the onus on them."[11]

In practice, however, as has been seen, the law operates by target setting when it can (it is easier to enforce), but the "reasonably practicable" approach may be used for deciding how far to go between upper and lower risk levels.

VI. THE COST OF SAVING A LIFE

In this section, the variations in the money spent to save a life are discussed. The extent to which it might be used as a basis for transfering expenditure from one activity to another is also considered. Various methods have been suggested for estimating the value of a life; however, the value of anything is the test of the marketplace — what people are prepared to pay for it — and that is the value that is discussed here. In some cases the valuation is explicit. Road engineers and those who authorize expenditure on road improvement schemes often use the costs as a basis for their decisions. Most of the estimates, however, are implicit. People do not know how much money they are spending per life saved as they have never made the necessary calculations. Stork writes " . . . government agencies as well as engineers in industry often strive for perfection, often without realizing what the cost of this perfection might be."[12]

Table 1 lists some estimates that have been published from 1967 onward, corrected to 1985 prices. No great accuracy was claimed for the figures and the earlier ones may have been made out-of-date by changes in technology, but they cover such a wide range, over six orders of magnitude, that such errors are relatively unimportant.

The last entry in the table may require some explanation. The probability of death per year for a member of social class 5 (unskilled occupations) in the U.K. is 1.8 times that of a member of social class 1 (professional occupations). If we assume that a social class 1 income will produce a social class 1 probability of death per year, then one can calculate the cost per life saved from the cost of raising the incomes of members of social class 5 to that of members of social class 1.[13]

Of course, it is not as simple as this. In the short-term, if more money was given to the poor, some of them might spend it on drinking or smoking and increase their probability of death per year. However, in the long run, rises in the standard of living of populations or social groups have been accompanied by falls in the probability of death per year.

If instead of giving members of social class 5 a social class 1 income, as assumed in Table 1, we give them a social class 2 income, then the cost per life saved would be halved, as members of social class 2 have almost the same probability of death per year as members of social class 1, but lower incomes.

Looking at the figures in Table 1, it is doubtful if there is any other commodity or service for which the market price varies so much. What would we say if the cost of a loaf of bread was 1 cent in one shop and over $10,000 in another across the road? (Electricity from watch batteries cost 10^5 times as much as electricity from the utility companies, but we pay for the convenience.)

The variations in the money spent to save a life, or some of them, might be justified if society had adopted a "target-setting" approach and had decided to try to reduce all risks to life to much the same level. In fact, this has not occurred. Most of the money seems to be spent in areas where the risks are already lowest.

It is unrealistic to expect precise equality in the money spent to save a life for several reasons:

1.　Some life-saving activities prolong life longer than others. For this reason some authors quote a cost per extra year of life rather than a cost per life saved.
2.　Society may judge that some forms of dying are more unpleasant than others and that more money should be spent to prevent them.
3.　Society may judge that more should be spent to avoid accidents that kill many people at a time than those that kill them one at a time.
4.　Some life-saving activities produce more additional benefits, in prevention of injury and ill-health, than others.
5.　If the chemical industry, for instance, spent less money on safety there is no social mechanism by which the money saved could be used to reduce risks from, for example, agriculture or smoking.
6.　Society does not advance by uniform improvement on a broad front but by spear-heading; by showing what is possible in one industry or activity, others may be encouraged to follow.

For these reasons, there would be no worry if the money spent to save a life varied over a range of ten to one, perhaps more, but 1 million to 1 is a different story.

Many of the activities listed in Table 1 are under the control of governments; in the U.K., the government controls the allocation of resources to the health service, road improvements, and the nuclear industry, but the implicit life valuations in these three areas are widely different. The government is also responsible to some extent for the redistribution of income

Table 1
SOME ESTIMATES OF THE MONEY SPENT (OR PROPOSED TO BE SPENT) TO SAVE A LIFE (CORRECTED TO 1985 VALUES)

Activity	Money spent to save a life (£)	Date of estimate	Ref.
Health			
Increasing tax on cigarettes (U.K.)	Negative[a]	1985	14
Antismoking propaganda (U.K.)	Small[b]	1985	14
Lung X-rays for old smokers (U.K.)	2,000	1973	15
Cervical cancer screening (U.K.)	6,000	1973	15
(U.S.)	50,000	1975	25
Breast cancer screening (U.K.)	13,000	1973	15
(U.S.)	150,000	1975	25
Artificial kidney (U.K.)	40,000	1973	15
Isotope-power heart (U.K.)	100,000	1973	15
Intensive care (U.K.)	20,000	1973	16
Mobile intensive care (U.S.)	60,000	1975	25
Liver transplant (U.S.)	100,000	1985	28
Reducing exposure to X-rays (U.K.)	9 million	1980	17
Road travel			
Mandatory use of safety belts (U.S.)	500	1982	18
30 other schemes (U.S.)	20,000—8 million	1982	18
18 schemes (U.S.)	40,000—800,000	1975	25
6 schemes (West Germany) (from use of rescue helicopters to fitting headrests)	100,000—1.5 million	1980	19
Various schemes (U.K.)	30,000—20 million	1967	20
Those actually implemented (U.K.)	Up to 1 million	1967	20
Industry (U.K.)			
Agriculture (employees)	10,000	1972	21
Rollover protection for tractors	400,000	1974	21
Steel handling (employees)	1 million	1974	21
Pharmaceuticals (employees)	20 million	1974	21
Pharmaceuticals (users)	50,000	1974	21
Chemical industry			
Typical figure	4 million	1975	20
11 proposals on a particular plant	500,000—100 million	1972	20
Nuclear industry			
Typical employees and public (U.K.)	15—30 million	1980	17
Reducing discharges to sea at Sellafield (U.K.)	15 million	1980	17
Marginal expenditure (U.S.)	175 million	1980	22
Marginal expenditure (U.S.)			
Effluent treatment	10 million	1979	23
Preventing H_2 explosions	3 billion	1979	23
National Radiological Protection Board Proposals (U.K.)			
High doses (£ 500/man/rem)	7 million	1981	17
Low doses (£ 20/man/rem)	270,000	1981	17
Domestic			
Smoke alarms in houses (U.K.)	500,000	1980	17
Use of subway (U.K.)	100,000—750,000	1974	24
Safer smoking materials (U.K.)			
Rates of pay for dangerous jobs (U.K.)			

Table 1 (continued)
SOME ESTIMATES OF THE MONEY SPENT (OR PROPOSED TO BE SPENT) TO SAVE A LIFE (CORRECTED TO 1985 VALUES)

Activity	Money spent to save a life (£)	Date of estimate	Ref.
Social policy			
Preventing collapse of high-rise flats (U.K.)	100 million	1974	20
Giving members of social class 5 a social class 1 income (U.K.)			
Family of 4 young people	2 million	1972	See
Family of 2 old people	100,000	1972	text
Giving members of social class 5 a social class 2 income (U.K.)			
Family of 4 young people	1 million	1972	See
Family of 2 old people	50,000	1972	text
Third World starvation relief	10,000	1980	17
Immunization (Indonesia)	100	1975	25

Note: Other estimates are reported in References 26 and 27.

[a]   A 10% increase in tax reduces smoking by about 5% so there is a net increase in government revenue and lives are saved.

[b]   If we spend £ 10 million on antismoking propaganda and as a result 1000 people (less than 1 smoker in 10,000) stop smoking, then the cost of saving a life is about £10,000. Parainflation factors used: 1967, 6.00; 1972, 4.33; 1973, 4.00; 1974, 3.65; 1975, 3.04; 1979, 1.83; 1980, 1.50; 1981, 1.35; 1982, 1.15; and 1985, 1.00. Figures up to 1982 are taken from *The Daily Telegraph*, September 5, 1982. 1982/5 estimated as 1.15. U.S. dollars have been converted to pounds sterling at the rate of $1.5/£ as this corresponds to their purchasing powers.

by taxation and social services. At present it collects taxes from social class 5 (mainly value added tax (VAT), and taxes on alcohol and tobacco) and spends it on reducing, by minute amounts, the already low risks from the nuclear industry. In the long run the money would save more lives, perhaps ten times as many, if it could be left with the poor and it was used to improve their standard of living.

Is it unreasonable to hope that government will pay at least a little attention to value for money when spending other people's money? Government might reply that in a democracy, legislation affects what the public wants and the public wants the risks from nuclear energy reduced, but is not worried about deaths due to poverty, road accidents, or smoking.

It is true that in a democracy, government is (or should be) the servant of the public and ultimately must do as the public wants, but a good servant should not obey his master uncritically; he should first point out the consequences of any proposed action. This governments have failed to do.*

---

*   Since this chapter was written, the U.K. Health and Safety Executive has proposed that the control and recommended limits described at the end of the section on *Target Setting* are replaced by Maximum Exposure Limits and Occupational Exposure Standards (OES). The OESs would be set so that "nearly all workers may be repeatedly exposed day after day without adverse effects". The trades unions have proposed instead that health-based limits should be set by a committee of scientists concerned solely with the protection of health and who would not take into account other factors, such as how technically achievable the limits might be. Reasonably Practicable Limits, which must be achieved, would take present conditions and achievable objectives into account. (*Health and Safety at Work*, 9(1), 39, 1987.) These various proposals show that the concept of two targets is becoming increasingly popular.

ACKNOWLEDGMENTS

Thanks are due to the many colleagues who suggested ideas for this chapter or commented on the draft, and to the U.K. Science and Engineering Research Council for financial support.

REFERENCES

1. **Kletz, T. A.,** *Inst. Chem. Eng., Symp. Ser.* 34, 75, 1971.
2. **Kletz, T. A., Hazop and Hazan** — Notes on the Identification and Assessment of Hazards, Institution of Chemical Engineers, Rugby, U.K., 1983.
3. **Lees, F. P.,** *Loss Prevention in the Process Industries,* Vol. 1, Butterworths, London, 1980, chap. 9.
4. **Kletz, T. A.,** *Reliability Eng.,* 3(4), 325, 1982.
5. **Dunster, H. J. and Vinck, W.,** *Nucl. Eng. Int.,* p. 23, August 1979.
6. Health and Safety Executive, Guidance Note EH15, Threshold Limit Values, issued annually from about 1958 to 1980, but the early issues had a different title, Her Majesty's Stationery Office, London. This publication was a reprint of the list published by the American Conference of Government Industrial Hygienists.
7. Health and Safety Executives Guidance Note EH40/84, Occupational Exposure Limits 1984, Her Majesty's Stationery Office, London, 1984.
8. **Grant, S., (Chairman),** Report of the Asbestos Working Group, Health and Safety Executive, London, 1983.
9. **Grant, S.,** *Health Safety Work,* 6(16), 28, 1984.
10. *Health and Safety Monitor,* p. 1, August 1984. Reprinted in *Chem. Safety Summary,* 55(219), 74, 1984.
11. **Fife, I. and Machin, E. A.,** *Redgrave's Health and Safety in Factories,* 2nd ed., Butterworth's, London, 1982, 16.
12. **Stork, W.,** *Automot. Eng.,* 51(3), 32, 1973.
13. **Wildavsky, A.,** *Public Interest,* p. 23, Summer 1980.
14. **Peto, R. and Doll, R.,** *New Sci.,* No. 1440, p. 26, 1985.
15. **Leach, G.,** *The Biocrats,* Penguin Books, Harmondsworth, U.K., 1972, 11.
16. **Miller, H.,** *Medicine and Society,* Oxford University Press, London, 1973, 63.
17. **Roberts, L. E. J.,** *Nuclear Power and Public Responsibility,* Cambridge University Press, London, 1984, 78.
18. **McCarthy, R. L., Taylor, R. K., and Finnegan, J. P.,** Catastrophic events, actual risk versis societal impact, *Proc. Annual Reliability and Maintainability Symp.,* The Forum for the Assurance Technologies, 1982.
19. **Dinman, B. D.,** *JAMA,* 244(11), 1226, 1980.
20. **Kletz, T. A.,** *Proceedings of the World Congress of Chemical Engineering, Chemical Engineering in a Changing World,* Koetsier, W. T., Ed., Elsevier, Amsterdam, 1976, 397.
21. **Sinclair, C.,** *Innovation and Human Risk,* Centre for the Study of Industrial Innovation, London, 1972.
22. **Siddall, E.,** *Nucl. Safety,* 21, 451, 1980.
23. **O'Donnell, E. P. and Mauro, J. J.,** *Nucl. Safety,* 20, 525, 1979.
24. **Melinek, S. J.,** *Accid. Anal. Prev.,* 6, 103, 1974.
25. **Cohen, B. L.,** *Health Phys.,* 38(1), 33, 1980.
26. **Siddall, E.,** A Rational Approach to Public Safety, Conference on Health Effects of Energy Production, Chalk River, Ontario, Canada, September 1979.
27. **Siddall, E.,** Nuclear Safety in Perspective, Canadian Nuclear Association 19th Annual Conference, Toronto, June 1979.
28. **Fineberg, H. V.,** *Technol. Rev.,* p. 17, January, 1985.

Chapter 3

FAULT TREE ANALYSIS FOR PROCESS PLANTS

Ulrich Hauptmanns

TABLE OF CONTENTS

I. INTRODUCTION

In striving for the improvement of the safety of technical systems* numerous formalized procedures for plant safety analysis have been developed. Some of them are applied to process plants, as discussed in References 1 to 3. An account with special emphasis on their use in the petrochemical industry is given in Reference 4. Most of the methods such as Failure Mode and Effects Analysis (FMEA) or Hazard and Operability studies (HAZOP) are qualitative and normally used on their own. They may be applied, however, in preparation of a fault tree analysis, a method usually employed for calculating the probability or expected frequency of an undesired event in a technical system.[5,6] If safety is the concern, in the case of process plants, this event may be an explosion, a fire, or a toxic release. If economic plant operation is the objective, it may simply be an unplanned outage. In this chapter, the concern is with fault tree analysis and its application to process plant safety.

Fault tree analysis is widely used in the nuclear power industry and constitutes the core of the methodology applied in risk studies,[7] where in addition to calculating the expected frequency of undesired events, their consequences are assessed.[8,9] Relatively few applications of the method to process plant systems have become known. They mostly refer to trip systems of hazardous installations,[10] auxiliary systems, or parts of chemical processes.[11-15] In contrast with the situation in the nuclear industry where results from fault tree analyses and risk studies are used to support the licensing process, the object of the principal publicly known risk study for process plants, which makes wider use of fault tree analysis, has been to explore the feasibility of the methodology in the field.[16] It is considered to be a useful approach.

The fault tree analyses for process plant systems, which have become known, deal with events during normal steady-state operation. Start-up and shut-down, which frequently give rise to accidents, are not addressed.

II. BASIC IDEAS

Fault tree analysis is a deductive method which is normally used in a quantitative way, although it requires as an initial step a qualitative study of the system under consideration, just as any method of systems analysis. After defining the undesired event, its logical connections with the basic events** of the system are searched for and the result of this search is represented graphically by means of a fault tree, as, for example, in Figure 1. The tree reflects the outcome of the qualitative part of the analysis, in which questions of the

---

* The term system is used to denote the object of analysis, which may be either part of a plant or an entire plant.

** The term "basic event" is understood to refer to the failure of technical components, the failure to carry out a human intervention, and the occurrence of external events like e.g., flooding of the plant or impact from neighboring installations.

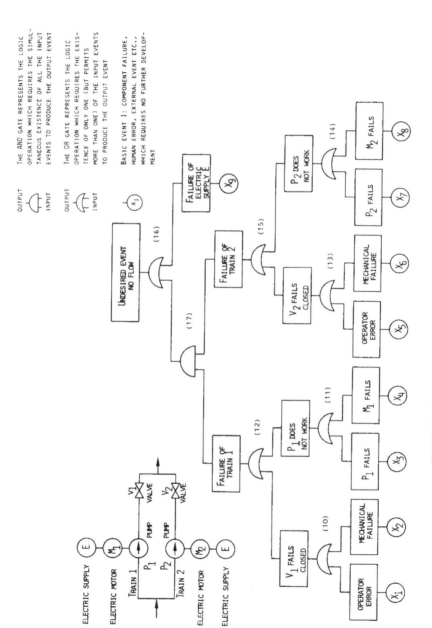

FIGURE 1. Fault tree for a system for transporting a fluid.

"how can it happen?" type are answered. These serve to identify, firstly, process functions and subsystems such as cooling or electric supply whose failure causes the undesired event and then connects these failures successively with the basic events.

The logical connections in the fault tree are generally represented by two types of gates, the "OR" and the "AND". In the case of the "OR" gate, any one of the entries alone is capable of producing the output event, while the output event of an "AND" gate only occurs if all its entry events are fulfilled. Sometimes in addition, "NOT" gates are used which convert the entry into its opposite. Two states of the basic events are normally admitted. They are either true or false* (e.g., failed or functioning in the case of technical components), which implies two possible states for the undesired event, its occurrence, and its nonoccurrence. The two states are adopted with certain probabilities which in the case of technical components are generally obtained for each type of component by evaluating the operating behavior of a great number of similar components. Applying these probabilities to the basic events of the fault tree, the probability of the undesired event may be calculated.

Since the functioning or failure of a technical system is determined by the values of its physical and chemical parameters such as temperatures, pressures, mass flows, concentrations, etc., the fault tree is a simplified representation of the system. It has the advantage of allowing to take into account the interrelationship of all the components of the system and the effect which their failure has on it. To model a system with consideration to all the above-mentioned parameters and their dynamic evolution as a consequence of a basic event, on the other hand, would be too complex a task. Nevertheless, information about the dynamic behavior of the physical and chemical parameters of the system as a result of basic events is required for working out the fault tree. This information is reflected by the logical structure of the tree where decisions such as whether a temperature will be reached which exceeds material limits, are underlying. Knowledge about system behavior is usually obtained from dynamic calculations of loads on individual components, experiments, or, in their absence, from engineering judgement. The latter should be exercised in a conservative way in safety analyses, i.e., the worst outcome for the system should always be assumed. It is normally supposed in fault tree analysis that components are designed in such a way that they fulfill their mission if they work correctly. For example, a relief valve is assumed to have a cross section and geometry such that the pressure in the area it protects remains below specified limits when relief is required. The assumption that components fulfill their mission has to be checked and special care must be taken, however, if failure combinations are contemplated in the fault tree which lead to loads beyond those on which component design was based. For then correct functioning may not always be assumed.

Fault tree analysis may be used in the design stage of a plant and thus influence its definite layout or may be employed to analyze already existing plants and provide an idea of the efficiency of traditional design procedures for safety and hints for plant improvement. Its application is particularly indicated when little historical evidence on a particular type of plant is available; in that case its probability of accident cannot be obtained directly from statistical records, but only by relating it with the failure of its components and other basic events as described earlier. At this point, advantage may be taken of the fact that the same type of component is usually encountered several times in a plant and that different types of plants normally use the same type of components. In addition, components generally fail more frequently than systems. This allows sufficient operating experience (number of components × time of operation) to be gathered in relatively short periods of time, in order to obtain a basis for estimating component failure probabilities. In similar ways probabilities

---

* Recently, extensions to more than two states have become known;[17,18] however, these have not attained practical relevance, since it is difficult to obtain probabilities for intermediate component states, e.g., partially opened valves, and it is usually impossible to predict system behavior in the case of such a type of failure.

for other basic events may be established. Although it may become possible with an increase of operating experience and the number of plants of the same type to estimate the probability of accident for a specific type of plant from recorded incidents, fault tree analysis would not lose its value since it provides insight into system structure, permits the identification of weak points, and allows the assessment of the influence of remedial measures. This type of information may normally not be derived from statistics about a particular type of plant. Generally speaking, the fault tree approach, which is based on the search for circumstances which make the system fail, may be regarded as an antithesis of the design process which strives to establish the conditions of system functioning. This is why it proves particularly useful in detecting design flaws both at the qualitative and quantitative levels of the investigation.

In order to carry out a fault tree analysis the following steps are required:

1. Familiarization with the system using process descriptions, piping and instrumentation diagrams, etc., and information obtained from the plant personnel
2. Definition of the undesired and initiating events using material information, checklists, historical evidence, etc.
3. Development of the fault tree(s)
4. Obtaining probabilities for the failure of technical components and human error
5. Evaluation of the fault tree(s)
6. Analysis of the results, proposals for system improvement, if necessary, corresponding changes of the fault tree(s), and renewed fault tree evaluation

The first step is very much a matter of organization and will not be treated here. The interested reader is referred to Reference 7. The remaining tasks, however, are discussed in detail in the following sections.

III. IDENTIFICATION OF UNDESIRED AND INITIATING EVENTS

A. General Remarks

Chemical processes involve both physical and chemical hazards. Physical hazards derive from operating conditions which may be extreme, such as very low or very high temperatures and pressures. Chemical hazards are those associated with the materials present in the process, which may be toxic, flammable or explosive, or exhibit several of these properties at the same time. The matter is complicated further by the fact that some of these properties may vary with changes of process parameters such as temperatures, pressures, or concentrations, or that these changes may give rise to side or spontaneous reactions,* for example, heating, decomposition, or polymerization. As it happens, incidents in chemical plants are characterized by these changes only. In addition, dangerous properties, if not present under normal process conditions, may evolve upon contact of process media with auxiliary media such as coolants, lubricants, or impurities, which may be introduced into the process with process streams or stem from component materials. After release they may occur as a consequence of reactions with substances present in the environment. The above enumeration, which is by no means complete, shows what difficulties the process plant safety analyst has to face as compared with his colleague in charge of nuclear reactors. The latter usually deals with well-defined situations, for example, that there should be no imbalance between heat production and removal from the reactor core such that cooling is no longer adequate.

---

* An event or reaction is considered as spontaneous in the present context, if they occur with a certain probability without the reason for their occurrence being totally known.

B. Undesired Events

The undesired event in a safety analysis for a chemical process plant usually is a toxic release, a fire or an explosion, or a situation in which these may be produced as, for example, pressure build-up due to a runaway reaction or the release of a substance which may evolve dangerous properties upon contact with the environment. Conservatively, it is often assumed that apart from air, water and ignition sources are present in the environment. Once the undesired event or events have been fixed (because several of them may occur at the same time), the initiating events — events potentially capable of bringing about the undesired event — must be found. This is the more complicated task of the two.

C. Initiating Events

If a system is designed properly, incidents can only occur if there are deviations from normal operating conditions. These may be provoked by component failures, which imply either the loss of a function (stuck valve, for example) or the loss of integrity (e.g., untight seal) and spontaneous/external events or human error.

Systems usually have components which are required to be working in order for the system to function (operational components), stand-by components which take over from them should they fail, and components which belong to protection and safety systems, and hence only have to work under special circumstances. Since only the failures of the operational components may affect system behavior directly, they are usually taken to be initiating events. In addition, the loss of component integrity, e.g., of the system boundary, in such a way that a release from the system or the ingress of air or auxiliary media become possible has to be considered, if dangerous situations may result. Human error must be contemplated as well. As mentioned earlier, some substances are capable of spontaneous reactions such as heating, decomposition, or polymerization. If such a reaction is possible in the system under investigation, it should be included in the list of initiating events. Furthermore, external events (e.g., aircraft crashes, tornadoes, flooding, blast waves, possibly from incidents in neighboring plants), have to be taken into account, if the scope of the analysis requires it.

D. Outline of a Computer-Aided Search for Undesired and Initiating Events

The search for undesired and initiating events requires the analyst to have a thorough knowledge of the system under investigation and a good background in physics, chemistry, and engineering. His ability for detecting dangerous situations should be enhanced by experience with previous analyses and an overview of past incidents in the same type or similar plants. This knowledge, together with specific information on the properties of the materials involved, the process conditions, and component failure modes (e.g., a valve may fail open or closed, leak, or be stuck), is combined to identify the undesired and initiating events.

This situation lends itself to building an expert system.[19] Therefore, a preliminary computer program was written which combines material properties from References 20 and 21, information on possible failure modes of components obtained in the field study,[22] and case histories from References 21 and 23 with input information on the process to yield specific warnings and undesired and initiating events, as shown in Figure 2. The program processes information using rules of the form IF . . . THEN, and it is assumed in line with the spirit of a safety analysis that anything that might happen will happen: IF the initial condition is satisfied THEN the outcome will occur with probability 1. For the case histories, a screening process according to material and event type is carried out. The user must then select those events which proceed in his specific case. Operational component failures are simply input after consulting a list with possible failure modes. The results of an analysis performed with the program are recorded and provide feedback in form of a checklist for later use. The application of the program is shown in the following section.

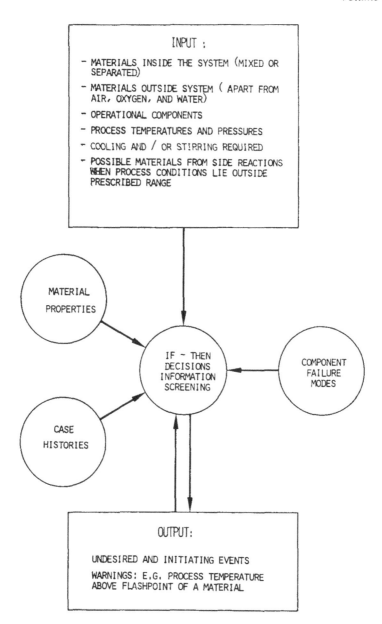

FIGURE 2. Schematic representation of a computer aided system for the identification of undesired and initiating events.

E. Example: Identification of Undesired and Initiating Events for the Ammonia-Air Mixing System of a Plant for the Production of Nitric Acid

The plant, whose simplified piping and instrumentation diagram is shown in Figure 3, serves to burn ammonia with the oxygen of the air to form nitrogen monoxide. It is discussed in detail in Reference 14. The system disposes of a pneumatic control system which maintains the ammonia-air ratio between 9 and 12% in volume by acting on valve PV1. Should this ratio lie outside the range between 8 and 13%, its electric shut-down system would activate solenoid valves SV1 and SV2, which in turn would close PV1 and PV2 and thus cut off the ammonia supply to the venturi jet E where the mixing with the air takes place. An additional safety feature is the temperature switch TS which activates SV1 and SV2, if the reaction temperature exceeds 910°C.

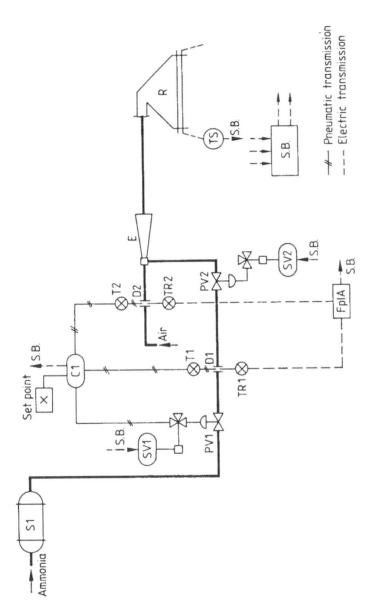

FIGURE 3.   Simplified piping and instrumentation diagram for the ammonia-air mixture and combustion system.

To find the undesired and initiating events, the computer program outlined in the previous section requires the following inputs:

- Materials inside system? (ammonia, air)
- Materials outside system? (default value: air, oxygen, water)
- Ammonia separated from other partners of the reaction? (no)
- Events outside system desired? (no)
- Minimum pressure? (>atmospheric)
- Failure of operational components? (PV1, C1, T1, T2)
- Human error? (yes, setpoint wrong)
- Consideration of external effects? (no)
- Search for relevant case histories? (yes)
- Initiating event from case history? (yes, formation of dangerous spray as described in Reference 23)
- Maximum and minimum permissible temperatures, pressures, concentrations, side reactions possible if the aforementioned parameters lie outside their ranges? (no)
- Type of reaction? (exothermic), cooling required? (no), stirring required? (no)

It then yields the following results:

Undesired event: ammonia forms an explosive mixture with air
Initiating events:

- Failure of PV1 closed
- Failure of C1
- Failure of T1 giving low signal
- Failure of T2 giving high signal
- Setpoint error
- Failure of heat exchanger S1 with formation of dangerous spray (from case history in Reference 23)

If possible undesired events outside the system had been asked for, toxic release, fire, and explosion owing to ammonia would have been indicated as undesired events and the failure of the pressure retaining boundary as an initiating event in addition to those given above.

IV. FAULT TREE DEVELOPMENT

A technical system usually disposes of a number of stand-by components (which may step in if the operational components fail) and protective and safety systems. These are normally capable of coping with the major part of initiating events and may be considered as barriers between those and the undesired event. The latter only occurs if these barriers fail. The situation is shown schematically in Figure 4. As can be seen, for some initiating events several barriers exist, while others directly provoke the undesired event. The number of barriers depends on the number of stand-by components and protective and safety systems available in the case of each particular initiating event.

If components from several barriers have to fail for the undesired event to occur, these are combined with the initiating event by an "AND" gate. If several of these combinations exist, they are input into an "OR" gate, just as the contributions from the different initiating events to the undesired event. The components which have to be in failed state at the same time if the initiating event is to cause the undesired event are called redundancies and their number indicates the degree of redundancy.

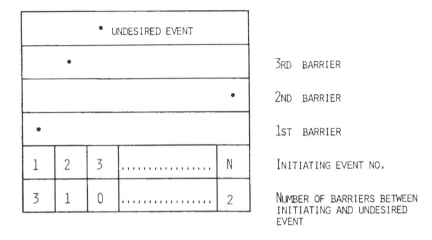

FIGURE 4.   Barriers against the occurrence of an undesired event

According to the lines described above, fault trees are developed. In the case of the system of Figure 3 this gives the tree presented in Figure 5.

V. ACQUISITION OF PROBABILITIES FOR THE FAILURE OF TECHNICAL COMPONENTS AND HUMAN ERROR

A. Mathematical Description of Component Behavior

The probability with which a component adopts its two possible states in practical work is taken to be either a constant value or is described by an exponential distribution.

If the behavior of a component i is characterized by a constant probability, either its unavailability, u_i, i.e., the probability of its being in failed state, or its complementary value, the availability, $p_i = 1 - u_i$, are indicated. If the description is in terms of an exponential distribution, the corresponding probability density function is

$$f_i(t) = \frac{1}{T_i} \exp\left(-\frac{t}{T_i}\right) \qquad (t > 0)\ (T_i > 0) \qquad (1)$$

Equation 1 yields upon integration over time t the unreliablity, i.e., the probability that component i experiences its first failure until the instant of time t:

$$q_i(t) = 1 - \exp\left(-\frac{t}{T_i}\right) \qquad (t > 0)\ (T_i > 0) \qquad (2)$$

In Equations 1 and 2, T_i is the mean time to failure for components of type i. This parameter is the inverse of the frequently used failure rate λ, i.e., $\lambda_i = 1/T_i$, which gives the probability of failure in an infinitesimal increment of time under the condition that the component has not failed before. If the component in question is not an object of maintenance its unreliablity and unavailability coincide, i.e., $q_i(t) = u_i(t)$.

Constant failure probabilities are used for components which have to function on demand, such as interruptors, if their lifetime chiefly depends on the number of demands they have experienced. There seems to be an indication, however, that a description in terms of failure rates is to be preferred, since the lifetime of components of this type is apparently more strongly influenced by factors which depend on the period of installation (such as corrosion) than by the number of demands.[22] Other fields of application for constant failure probabilities

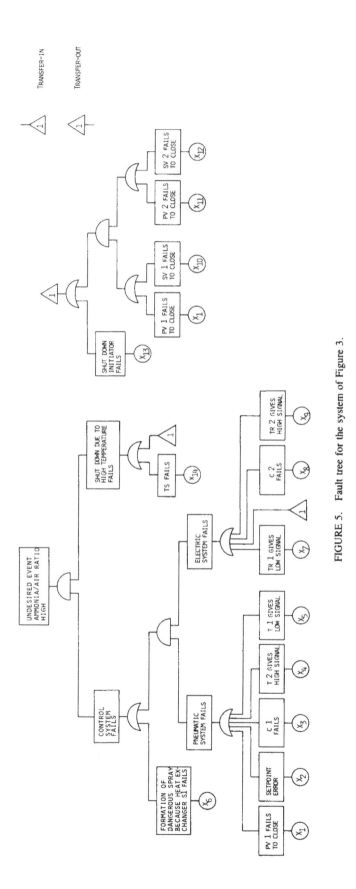

FIGURE 5. Fault tree for the system of Figure 3.

are the treatment of human error and of operational characteristics, e.g., the probability that a heat exchanger is not in its functioning phase. In addition, they may be used for components subjected to maintenance whose asymptotic unavailability is given by:

$$\frac{downtime}{downtime \ + \ functioning \ time} \tag{3}$$

In Equation 3 the downtime is the period during which the component is out of service, either because its failure has not yet been detected or because it is disconnected during its repair. In all other cases failure rates are generally used. If information is available, failure rates or probabilities are indicated for different modes of failure, each of which may figure as a separate basic event in the fault tree.

Both reliability parameters are obtained from experience by statistically evaluating the number of failures per demand or time of operation. Generally several components of the same type or groups of sufficiently similar components are considered as representative of a certain class of component. Their behavior is observed during a period of time and mean values and confidence bounds are calculated, assuming that their failures can be described by a binomial distribution (constant failure probabilities) or a Poisson distribution (failure rates). In order to be able to evaluate properly, an inventory of all the components of the system where the observation takes place is made. The inventory contains among others the technical characteristics of each type of component, (e.g., size, material, number of revolutions, and power) and parameters characterizing its conditions of operation, (e.g., type of medium, temperature, and pressure.) Only in this way can an adequate grouping of components be achieved and may possible dependencies of lifetimes on these parameters be detected.

Failure rates are generally supposed to exhibit a time behavior which can be described by the so-called "bathtub curve". At the beginning of component lifetime failures are relatively frequent (burn-in period). After that follows a time interval with a virtually constant failure rate. Toward the end of the lifetime another increase due to aging can be observed.[24] The same is true for components characterized by unavailabilities on demand. Components are generally installed after their burn-in period, and owing to periodic inspections replaced before aging becomes a major problem. Therefore, time-independent failure rates* and unavailabilities on demand are normally used in fault tree analysis. This practice is to a certain extent supported by experience, since a long time observation performed in a nuclear power station has not revealed any time dependence.[22]

In addition to the quality of the components maintenance has an influence on system performance. Two basic types of maintenance may be distinguished: preventive and corrective. In preventive maintenance, periodic inspections are carried out which are meant to discover anomalies which have not yet led to a failure and remedy them before a failure occurs. Corrective maintenance implies repair or substitution of the component after its failure has occurred. In order to be introduced into fault tree analysis maintenance has to be described mathematically. This may be achieved using the theory of Markov processes or renewal theory.[25,26] No details of these procedures will be given here, however. Instead, a simple frequently used mathematical formulation for preventive maintenance is presented.[27] As far as corrective maintenance is concerned reference is made to Equation 3.

The model for periodic inspection is valid under the following conditions:

- The lifetime of the component may be described by an exponential distribution
- The time between inspections is constant throughout component lifetime

---

* Constant, i.e., time-independent failure rates of necessity lead to the exponential distribution of Equation 2.[24]

- Failures are only detected on the occasion of an inspection
- The duration of the repair is negligible compared with the mean time to failure of the component
- After inspection the component is assumed to be "as good as new"

The unavailability of the component is then given by:

$$u_i(t) = 1 - \exp[-\lambda_i(t - n\,\theta_i)] \qquad (t > 0) \qquad (4)$$

In Equation 4, λ_i is the failure rate of component i, θ_i is the time between inspections, and n is the integer part of the quotient t/θ_i. It implies that the time dependence of component unavailability is identical between any two inspections.

Frequently, the average unavailability is used to characterize a component. This parameter is obtained after integrating Equation 4 over the time interval between two inspections:

$$\bar{u}_i = 1 + \frac{T_i}{\theta_i}[\exp(-\lambda_i\theta_i) - 1] \qquad (5)$$

Equation 5 may be further simplified, if $\lambda_i\,\theta_i \ll 1$. In this case the exponential function may be approximated by the first three terms of its Taylor series. This procedure leads to the well-known result:

$$\bar{u}_i \cong \frac{\lambda_i\,\theta_i}{2} \qquad (6)$$

B. Reliability Data for Process Plant Analyses

The reliability data used in fault tree analyses should ideally have been obtained in a system which is similar to that under investigation, i.e., similar components work under similar operating conditions. This goal may be satisfied in the case of specific types of nuclear reactors using the data compiled in Reference 22. Comprehensive field studies for process plants, on the other hand, have not become known. The data proceeding from the literature[28,29] do not supply component characteristics and give hardly any indication as to operating conditions. The knowledge of both is essential for making an adequate choice of reliability data in fault tree analysis. The only way out of this dilemma, at present, is the use of environmental factors with which unavailabilities and failure rates are multiplied in order to take into account special operating conditions (factors <1 for very light requirements, >1 if conditions are more taxing than "normal"). The procedure naturally is very approximate and hence unsatisfactory. Some data obtained in this way are communicated in the last section.

An additional problem in fault tree quantification derives from the fact that probabilities for the occurrence of spontaneous chemical reactions are hardly ever known and therefore have to be estimated.

C. Common Mode Failures

Apart from the independent failures treated previously, the possibility of common mode failures in technical systems has to be contemplated. This type of failure leads to the simultaneous unavailability of several components and is especially grave if it affects a redundant structure, i.e., several components instead of one arranged in such a way that any one of them is capable of performing the mission of the principal component. In such a structure a common mode failure may ruin the increase in availability it provides.

The following types of common mode failures may be distinguished[9]:

1. Failures of two or more redundant components or partial systems which are of similar or identical design owing to an outside cause, for example, a corrosive environment which leads to rapid component degradation
2. Failures of two or more redundant components or partial systems which occur as a consequence of a single failure; this type of common mode failure is called causal failure
3. Failures of two or more redundant components or partial systems which occur as a consequence of functional dependencies as, for example, the dependence on a common auxiliary system

The latter type of failure may be treated in the fault tree by adequate modeling. For example, the electric supply in the system of Figure 1 is such a case. The same applies to causal failures such as damage by flying fragments, pipe whip, or humidity. Frequently, however, causal failures are rendered impossible by separating systems physically.

The remaining types of common mode failures due to a common external cause (e.g., planning, construction, or maintenance errors) should be treated by evaluating relevant operating experience. In this category, which comprises the common mode failures in the strictest sense of the word, a distinction should be drawn between those (1) which occur or are discovered on occasion of an incident, (2) those which are discovered on functional demand of the system either because of testing or an operational requirement, or (3) those which are self-annunciating (e.g., because the components affected are of the type which gives an alarm upon failure). Operating experience primarily supplies data for the last two types of common mode failures, while those occurring only on the occasion of an incident can in general only be discovered using analytical methods.

The evaluation of operating experience may be carried out with several models; among them are the β-factor method and the specialized Marshall-Olkin model. These models are discussed in detail in References 30 and 31. Their application requires operating experience which unfortunately is scarce at present. This is not surprising if one considers the fact that common mode failures are less frequent than independent failures and observation periods in field studies would have to be unduly long for many failures of this type to be observed. In addition, the cause of a common mode failure is normally removed after its occurrence. Common mode data for process plants have not become known.

Any kind of common mode failure is introduced into the fault tree as a basic event which coexists with other basic events representing the independent failures of the components affected.

D. Human Error

Thus far only the failures of technical components have been considered. Since technical systems, however advanced their level of automation, still rely on human intervention in some respects, a fault tree analysis would be incomplete if this aspect were neglected. In modern process plants direct operator control is unusual. Automatic controllers generally ensure that process parameters are maintained close to nominal levels, except perhaps for start-up and shut-down, when an increased degree of human intervention is normally required. The operator's job therefore usually consists of a number of intermittent activities such as[3]:

1. Operational tasks
   * Sequential control, starting pumps and motors, opening and closing valves, etc, during start-up, shut-down, and batch processing operations
   * More direct control of process parameters when control loops are not working
   * Monitoring the plant for correct operation (compared with expected performance)
   * Carrying out manual operations such as loading materials into hoppers and carrying out manually steered operations such as crane control

- Collecting and changing paper on chart recorders
- Completing plant production and operation log books
- Taking samples and operating instruments
- Alarm response and diagnosis of unusual plant conditions
- Reporting and following up equipment failures
2. Maintenance tasks
    - Adjusting manually controlled valves and pipe couplings for correct line up
    - Adjusting set points for control loops and valve positions

In the context of a fault tree analysis operator interventions should be introduced as basic events into the fault tree and ultimately be quantified. A quantification is, at present, only possible for the failure of an operator to carry out a planned intervention, e.g., opening a valve to increase coolant flow when the temperature gets too high. Unplanned acts (playing around with buttons or changing positions of valves because of absent-mindedness or with the intention of causing harm) cannot be quantified. Even if this limitation is accepted, human error quantification still remains less exact than the quantification of the failure of technical components. Therefore, it may be recommended to calculate bounds for system reliability, assuming on the one hand perfect human intervention (u = 0) and complete failure (u = 1) on the other.[15]

Human error is most frequently treated by the methods exposed in Reference 32. A human error is defined there as an act outside tolerance limits. It is evident that the permissible interval of tolerance depends on the type of human act in question and on the circumstances under which it is carried out. The definition has to be made by the analyst in the light of these aspects. Usually the following kinds of human error are distinguished:

Error of omission: Failure to perform a task or part of a task, e.g., a step
Error of commission: Performing a task or a step incorrectly
Extraneous act: Introducing some task or step which should not have been performed
Sequential error: Performing some task or step out of sequence
Timing error: Failure to perform a task or step within the allotted time or per-
                      forming them too early or too late

The basis for the evaluation of human error is the identification of the acts to be carried out and their analysis. Important parameters to be established are the moment of the interventions, the time available for their realization, the information at hand (instrument readings, knowledge of process behavior, computerized information supply, etc.), and the possibility of correction if the initially required intervention has not been carried out. Ergonomic and environmental aspects have to be considered as well. In addition, it is important to take into account possible dependencies of human acts. These may be due to factors such as elevated stress which would affect several consecutive acts realized by the same person or circumstances which would influence the action of two different persons trying to carry out the same act such as difficult access to the place of intervention.

The most widely used method for human error quantification is THERP (Technique for Human Error Prediction), which is discussed in detail in Reference 32. It is based on assigning error probabilities to simple tasks and breaking down more complicated tasks into simple ones, whose probabilities are combined according to the laws of probability in order to obtain the error probability of the complicated task. In addition, performance shaping factors (factors affecting these probabilities significantly) are taken into account by multiplying the base values, which apply to "normal" conditions with them. In Reference 32 a great number of such performance shaping factors are discussed. In the present context only a few of the more important ones are commented upon:

Ergonomic layout of the control room: An increase of failure probabilities is to be assumed if the arrangement, labeling, and design of the control mechanisms is such that error is enhanced. This may be the case, for example, if stereotypes* are violated, or if labeling of instruments and buttons is confusing or hardly legible.

Feedback through indications and alarms: The probability of human failure is reduced, if feedback through indications and alarms which render the detection of an error probable exists. The possibility of the discovery of an error is to be taken into account especially if the operator is warned immediately after committing it. This applies most of all if system response to the error is rapid.

Human redundancy: A further important way of detecting errors results from human redundancy, i.e., a decision or an act involves more than one person with adequate qualification. Redundancy is also assumed if a person's acts are controlled by another.

Psychical stress: Stress is a very important factor for human performance. If it is too low, i.e., work is of routine type and considered as boring, error becomes more probable. If stress is very high, on the other hand, error again becomes very probable, reaching the value 1 for very dangerous situations. This value should be adopted, for example, for interventions immediately after a loss of coolant accident in a nuclear power station or during a runaway reaction, if it implies getting close to the reactor. Between these two extremes, there lies an optimal stress range which is assumed for control room work during normal operation, maintenance work, and testing.

Qualification and training of operators: Among other factors, appropriate qualification for the work to be done is essential for avoiding errors. This implies neither under- nor overqualification and includes general education and a specific understanding of the bases and procedures of the process in question. Another important aspect in this context is the training for emergency situations which helps to maintain an acceptable level of emergency response probability. This probability would otherwise decrease in the course of time. For this reason simulator training is required for aircraft pilots and nuclear reactor operators.

Written instructions: A good explanation of what should be done in operating the plant both in normal and emergency conditions in written form tends to reduce the probability of human error.

Human error is treated in fault tree analysis in analogy with the failure of technical components. Its quantification, however, is much more complicated than that of the latter and requires the collaboration of experts from various disciplines such as engineering, psychology, ergonomy, and statistics.

The application of the foregoing considerations in an analysis is shown in the last section.

E. Uncertainties

Uncertainties exist in the estimation of reliability data. In the case of technical components these may be due to: differences in the performance of components of the same class and grouping together of similar but not identical components working under similar but not identical operating conditions; if data from the literature is used to the necessity to select values from different sources without knowing whether component designs and operating conditions are comparable, and it is very probable that they are not. For this reason use of a statistical distribution for unavailabilities and failure rates is indicated instead of a single point value. Usually a lognormal distribution is chosen for this purpose because it fits observed data reasonably well.[22] The corresponding probability density function is in the case of λ (u is treated analogously)

---

*   A stereotype is the expected reaction of a human to an outside influence. For example, turning a button in a clockwise direction is associated with switching off.

$$f(\lambda) = \frac{1}{\sqrt{2\,\pi}\lambda s} \exp\left[-\frac{(\ln \lambda - \mu)^2}{2\,s^2}\right] \quad (\lambda, s > 0) \tag{7}$$

The parameters which appear in Equation 7 are calculated as follows:

$$\mu = \frac{1}{N} \sum_{n=1}^{N} \ln \lambda_n = \ln \lambda_{50} \tag{8}$$

and

$$s^2 = \frac{1}{N-1} \sum_{n=1}^{N} (\ln \lambda_n - \mu)^2 \tag{9}$$

Equations 8 and 9 are the mean and the variance of the natural logarithms of the failure rates, respectively. N is the total number of measured values or data taken from the literature and λ_{50} is the corresponding median. The mean value of the distribution is given by:

$$\lambda = \lambda_{50} \exp\left(\frac{s^2}{2}\right) \tag{10}$$

The dispersion of the data is usually characterized by indicating the factor:

$$K = \exp(s \cdot 1.6449) \tag{11}$$

The choice of the value 1.6449 in the argument of the exponential function in Equation 11 makes the probability of encountering a value of λ in the interval $\lambda_{05} = \lambda_{50}/K < \lambda < \lambda_{50} \cdot K = \lambda_{95}$ equal to 90%, with the probability of λ lying below or above these limits being 5% each. They therefore represent the 5 and 95% centiles of the distribution, respectively.

If insufficient data are available for calculating the dispersion factor K, an estimate is frequently made which reflects the analyst's subjective judgment as to the uncertainty of the failure rate value.

Uncertainties of the probabilities for human error and other basic events are treated in the same way in the context of the analysis.

VI. FAULT TREE EVALUATION

A. Description of Basic Events and Systems with Binary Variables

Any system represented by a fault tree has basic events which act in series ("OR" gates) or in parallel ("AND" gates), with a combination of the two being most frequent. The basic event may be described by a binary variable[*][24]:

$$x_i = \begin{cases} 1, \text{ if basic event i is true,} \\ \\ \text{e.g., component i has failed} \\ \\ 0, \text{ if basic event i is not true,} \\ \\ \text{e.g., component i is working} \end{cases} \tag{12}$$

* A binary variable only adopts two states, e.g., 0 and 1.

An analogous description may be used for the state of the system:

$$\psi = \begin{cases} 1, \text{ if the undesired event has occurred} \\ \\ 0, \text{ if the undesired event has not occurred} \end{cases} \quad (13)$$

Assuming, as usual, that the state of the system is completely described by the state of its basic events, then $\psi = \psi(\vec{x})$, where $\vec{x} = (x_i, \ldots, x_I)$, and I denotes its total number of basic events. $\psi(\vec{x})$ is called the structure function of the system. It is supposed to be monotonously nondecreasing, i.e., to describe systems which, after the occurrence of the undesired event, do not return to their original state if an additional basic event is true. This property is guaranteed if only "AND" and "OR" gates figure in the fault tree.

B. Minimal Cut Sets

A minimal cut set is a group of basic events which, in case they are true, are just sufficient to cause the undesired event, e.g., a group of components which are in failed state at the same time. In general, a fault tree has several minimal cut sets, each of which represents one way of bringing about the undesired event. Thus, for example, the fault tree of Figure 1 has the following minimal cut sets*:

$$K_1 = (9) \qquad K_2 = (1,5) \qquad K_3 = (1,6) \qquad K_4 = (1,7)$$

$$K_5 = (1,8) \qquad K_6 = (2,5) \qquad K_7 = (2,6) \qquad K_8 = (2,7)$$

$$K_9 = (2,8) \qquad K_{10} = (3,5) \qquad K_{11} = (3,6) \qquad K_{12} = (3,7)$$

$$K_{13} = (3,8) \qquad K_{14} = (4,5) \qquad K_{15} = (4,6) \qquad K_{16} = (4,7)$$

$$K_{17} = (4,8)$$

As can easily be seen, the simultaneous failure of the component(s) of any of the 17 combinations causes the flow through the system to be interrupted. On the other hand, if any of the components contained in a minimal cut set were to function again, the system would no longer be failed by this particular mode of failure.

The representation of the fault tree in terms of its minimal cut sets gives insight into the structure of the system under investigation. Thus, it can be seen which of the basic events alone or in conjunction with others can bring about the undesired event. Minimal cut sets with only one element represent a series structure, those with several elements a parallel structure (also called redundant). The aforementioned information provides the starting point for detecting weaknesses of the system design, since, for example, minimal cut sets with only one element signal the absence of redundancies and the appearance of the same element in several minimal cut sets shows that its failure contributes to various modes of system failure. Both situations should be regarded as potentially critical, unless the probability of the basic event being true is extremely low. The minimal cut sets allow the writing of the structure function of the system in the following way[24]:

$$\psi(\vec{x}) = 1 - \prod_{l=1}^{L} [1 - \kappa_l(\vec{x})] \quad (14)$$

---

* The numbers inside the parentheses denote the basic events which belong to the minimal cut set. They will occasionally be referred to as elements of the minimal cut set.

In Equation (14):

$$\kappa_i(\vec{x}) = \prod_{i \in K_i} x_i \tag{15}$$

The K_l represent the L minimal cut sets of the system, with the value of L depending on its complexity, and κ_l are the binary functions associated with them. These are known as minimal parallel cut structures. For example, the minimal cut set $K = \{2,4,7\}$ has the associated binary function $\kappa = x_2 \cdot x_4 \cdot x_7$, which is equal to 1 if basic events 2, 4, and 7 are true, and equal to 0, if any one or several of them are false. This result holds in general.

Fault trees for complex systems normally must be evaluated with the aid of computer programs. For this purpose, three principal types of methods are available: (1) direct simulation of the fault tree, (2) minimal cut set calculation using a simulation procedure, and (3) minimal cut set calculation by analytical methods. In what follows, these procedures will be discussed in more detail.

C. Evaluation of a Fault Tree by Direct Simulation

The evaluation of a fault tree by direct simulation is carried out using the Monte Carlo method.[33] The behavior of the components* is simulated in accordance with the distribution of their lifetimes. This is done by generating random numbers with a uniform distribution over the interval $0 < z_{i,j} < 1$, where the subscript denotes the jth generation of a random number for component i.[34]

If the component i is described by a constant unavailability, u_i, it is assumed failed, if:

$$u_i \geq z_{i,j} \qquad j = 1,\ldots, J \tag{16}$$

In Equation 16 I is the total number of components of the fault tree and J the total number of Monte Carolo trials chosen by the analyst. If, on the other hand, component lifetimes are distributed exponentially, the uniform random numbers are converted to this distribution giving:

$$r_{i,j} = -T_i \cdot \ln(1 - z_{i,j}) \qquad j = 1,\ldots, J \tag{17}$$

In Equation 17, $r_{i,j}$ is the lifetime of component i in trial j. The process of trials may be regarded as an inversion of the lifetime measurements for the component, which have led to the mean time to failure T_i. The lifetime $r_{i,j}$ is compared with the mission time $t = T_m$ for which the unreliability of the system is to be calculated. The component i is considered failed if $r_{i,j} < T_m$ and functioning otherwise.

The times until failure $r_{i,j}$ are arranged in ascending order beginning with their smallest value.** Then the logical function of the fault tree*** is consulted with all the components satisfying the condition of Equation 16 in failed state. Should the undesired event occur, this fact is recorded and the next trial, $j + 1$, is initiated. If, on the other hand, the undesired event does not occur, the component with the smallest time until failure is declared failed

---

* For ease of presentation, the description is made in terms of components. The application to other basic events is analogous.
** This is only necessary if maintenance is to be taken into account or minimal cut sets are to be determined, as described in the next section. If only reliability parameters are required it is sufficient to record if the basic events which are true cause the undesired event and then to proceed to the next trial.
***For example, $L = (x_1$ OR x_2 OR x_3 OR $x_4)$ AND $(x_5$ OR x_6 OR x_7 OR $x_8)$ OR x_9 in the case of the fault tree of Figure 1.

in addition to the other failed components, if its $r_{i,j} < T_m$, and the logical function is consulted once again. This process is continued until either the undesired event occurs or the component which would have to be declared as failed next has a time until failure which is greater than the mission time, T_m. After performing the total number of J trials the quotient of the number of occurrences of the undesired event and J gives the unreliability of the system for the mission time, T_m, $9_s (T_m)$.*

Since the process described is stochastic, the unreliability can only be indicated within certain confidence bounds which can be narrowed by increasing the number of trials. For a predetermined degree of precision the number of trials required rises with the inverse of the system unreliability, and may hence become prohibitive for very reliable systems. This drawback may be remedied to a certain extent using importance sampling, as described in References 35 and 36, where in addition the inclusion of maintenance schemes and the calculation of other reliability parameters are treated.

D. Determination of Minimal Cut Sets by Simulation

Minimal cut sets are determined applying the Monte Carlo method according to the scheme outlined in the previous section. Every time the logical function representing the fault tree signals that the undesired event has occurred, a group of components has failed. This group does not necessarily constitute a minimal cut set, because it may contain components whose failure is not necessary for causing the undesired event. These superfluous components have to be eliminated from the group so that the remainder becomes minimal.

Since in the present case it is not intended to calculate the unreliability directly, the necessity of a great number of trials for analyzing very reliable systems may be circumvented by choosing the mission time in such a way that the undesired event occurs in about half the trials.[37] This procedure makes use of the fact that a technical system without maintenance becomes more unreliable with increasing mission time.

With this method in general not all minimal cut sets are found, but only those which make a major contribution to system unreliability. Because of the stochastic nature of the process these are only obtained with a certain level of confidence, which may be increased by increasing the number of trials. The property of finding only part and not the totality of the minimal cut sets of a fault tree may be desired, if the tree has a great number of them (there may be several millions or more in some cases), all of which cannot be obtained because of obvious limitations of memory space and computing time.

E. Determination of Minimal Cut Sets by Analytical Methods

In analytical methods for evaluating fault trees use is made of Boolean algebra operations in order to transform the tree in such a way that it is expressed in terms of its minimal cut sets. In contrast with the preceding approach this procedure does not require reliablity data for obtaining the minimal cut sets of the tree, but only for calculating the probability of the undesired event. Hence, the process of obtaining the minimal cut sets is not affected by possible flaws in the data as may be the case if the Monte Carlo method is used for this purpose. Basically two approaches for evaluating a fault tree analytically exist: the "top-down" approach, in which the algorithm starts with the undesired event represented by the top gate working its way down to the basic events, and the "bottom-up" approach, where the calculation is initiated at the level of basic events and ends with the undesired event.

The procedure presented here is of the "top-down" type and is described in Reference 38. The fault tree of Figure 1 serves as an example to show how the algorithm works.

The tree is represented by a matrix in which the entry of a "1" indicates a connection

---

* Reliability parameters for systems (e.g., unavailability, unreliability) are defined as has been done for components in Section V.A.

and a "0" means that there is none. The rows of the matrix represent the "OR" gates (upper part) and "AND" gates (lower part). The columns are divided into three blocks. The first block contains the basic events of the tree, the second its "OR" gates, and the third its "AND" gates. Matrix $\overline{A}_o$, which is given below, is the representation of the fault tree of Figure 1 according to these conventions (the numeration of the gates is indicated in Figure 1).

	1	2	3	4	5	6	7	8	9		10	11	12	13	14	15	16	17
				Basic events							OR-gates							AND-
Gate no.																		gate
10	1	1	0	0	0	0	0	0	0		0	0	0	0	0	0	0	0
11	0	0	1	1	0	0	0	0	0		0	0	0	0	0	0	0	0
12	0	0	0	0	0	0	0	0	0		1	1	0	0	0	0	0	0
13 OR-	0	0	0	0	1	1	0	0	0		0	0	0	0	0	0	0	0
14 gates	0	0	0	0	0	0	1	1	0		0	0	0	0	0	0	0	0
15	0	0	0	0	0	0	0	0	0		0	0	0	1	1	0	0	0
16	0	0	0	0	0	0	0	0	1		0	0	0	0	0	0	0	1
17 AND-gate	0	0	0	0	0	0	0	0	0		0	0	1	0	0	1	0	0

$\overline{A}_o =$ (indicated to the left of the matrix)

Looking at matrix $\overline{A}_o$ the objective of the algorithm becomes clear: the transformation of the matrix into a form which only contains "0" in the two blocks on the right side, which represent the gates. This is achieved by replacing the gates systematically by their entries. In the case of an "AND" gate, all the entries figure in the original row, where the replacement is made. If a gate is of the "OR" type, for each of its entries a new row is opened. The practical realization of these operations takes place making use of the Boolean algebra operations for the logical "AND" and "OR". The algorithm starts with the top gate. In the present example this means replacing gate "16" in matrix $\overline{A}_o$ by its entries, which leads to:

	Basic events									Gates							
$\overline{A}_1 =$	0	0	0	0	0	0	0	0	1	0	0	0	0	0	0	0	0
	0	0	0	0	0	0	0	0	0	0	0	0	0	0	0	0	1

The first row of matrix $\overline{A}_1$ already contains a representation in terms of components (only component "9") and therefore is retained unchanged in further steps. Replacement of gate "17" in the second row, which represents an "ND" gate, leads to:

$$\overline{A}_2 = \begin{bmatrix} 0 & 0 & 0 & 0 & 0 & 0 & 0 & 0 & 1 & | & 0 & 0 & 0 & 0 & 0 & 0 & 0 & 0 \\ 0 & 0 & 0 & 0 & 0 & 0 & 0 & 0 & 0 & | & 0 & 0 & 1 & 0 & 0 & 1 & 0 & 0 \end{bmatrix}$$

$$\text{Basic events} \qquad \qquad \text{Gates}$$

In the next step gate "12" in the second row of $\overline{A}_2$ is replaced. It is of the "OR" type which implies that for each of its entries a new row is opened. This leads to:

$$\overline{A}_3 = \begin{bmatrix} 0 & 0 & 0 & 0 & 0 & 0 & 0 & 0 & 1 & | & 0 & 0 & 0 & 0 & 0 & 0 & 0 & 0 \\ 0 & 0 & 0 & 0 & 0 & 0 & 0 & 0 & 0 & | & 1 & 0 & 0 & 0 & 0 & 1 & 0 & 0 \\ 0 & 0 & 0 & 0 & 0 & 0 & 0 & 0 & 0 & | & 0 & 1 & 0 & 0 & 0 & 1 & 0 & 0 \end{bmatrix}$$

$$\text{Basic events} \qquad \qquad \text{Gates}$$

Replacing gates "10" in the second row of $\overline{A}_3$ and "11" in its third row the following representation is obtained:

$$\overline{A}_4 = \begin{bmatrix} 0 & 0 & 0 & 0 & 0 & 0 & 0 & 0 & 1 & | & 0 & 0 & 0 & 0 & 0 & 0 & 0 & 0 \\ 1 & 0 & 0 & 0 & 0 & 0 & 0 & 0 & 0 & | & 0 & 0 & 0 & 0 & 0 & 1 & 0 & 0 \\ 0 & 1 & 0 & 0 & 0 & 0 & 0 & 0 & 0 & | & 0 & 0 & 0 & 0 & 0 & 1 & 0 & 0 \\ 0 & 0 & 1 & 0 & 0 & 0 & 0 & 0 & 0 & | & 0 & 0 & 0 & 0 & 0 & 1 & 0 & 0 \\ 0 & 0 & 0 & 1 & 0 & 0 & 0 & 0 & 0 & | & 0 & 0 & 0 & 0 & 0 & 1 & 0 & 0 \end{bmatrix}$$

$$\text{Basic events} \qquad \qquad \text{Gates}$$

Transformation of matrix $\overline{A}_4$ along the same lines leads to:

$$\overline{A}_5 = \begin{bmatrix} 0 & 0 & 0 & 0 & 0 & 0 & 0 & 0 & 1 & | & 0 & 0 & 0 & 0 & 0 & 0 & 0 & 0 \\ 1 & 0 & 0 & 0 & 0 & 0 & 0 & 0 & 0 & | & 0 & 0 & 0 & 1 & 0 & 0 & 0 & 0 \\ 1 & 0 & 0 & 0 & 0 & 0 & 0 & 0 & 0 & | & 0 & 0 & 0 & 0 & 1 & 0 & 0 & 0 \\ 0 & 1 & 0 & 0 & 0 & 0 & 0 & 0 & 0 & | & 0 & 0 & 0 & 1 & 0 & 0 & 0 & 0 \\ 0 & 1 & 0 & 0 & 0 & 0 & 0 & 0 & 0 & | & 0 & 0 & 0 & 0 & 1 & 0 & 0 & 0 \\ 0 & 0 & 1 & 0 & 0 & 0 & 0 & 0 & 0 & | & 0 & 0 & 0 & 1 & 0 & 0 & 0 & 0 \\ 0 & 0 & 1 & 0 & 0 & 0 & 0 & 0 & 0 & | & 0 & 0 & 0 & 0 & 1 & 0 & 0 & 0 \\ 0 & 0 & 0 & 1 & 0 & 0 & 0 & 0 & 0 & | & 0 & 0 & 0 & 1 & 0 & 0 & 0 & 0 \\ 0 & 0 & 0 & 1 & 0 & 0 & 0 & 0 & 0 & | & 0 & 0 & 0 & 0 & 1 & 0 & 0 & 0 \end{bmatrix}$$

$$\text{Basic events} \qquad \qquad \text{Gates}$$

Finally,

$$\overline{A}_6 = \begin{bmatrix}
 & 1 & 2 & 3 & 4 & 5 & 6 & 7 & 8 & 9 & & 10 & 11 & 12 & 13 & 14 & 15 & 16 & 17 \\
 & \multicolumn{9}{c}{\text{Basic events}} & & \multicolumn{8}{c}{\text{Gates}} \\
 & 0 & 0 & 0 & 0 & 0 & 0 & 0 & 0 & 1 & | & 0 & 0 & 0 & 0 & 0 & 0 & 0 & 0 \\
 & 1 & 0 & 0 & 0 & 1 & 0 & 0 & 0 & 0 & | & 0 & 0 & 0 & 0 & 0 & 0 & 0 & 0 \\
 & 1 & 0 & 0 & 0 & 0 & 1 & 0 & 0 & 0 & | & 0 & 0 & 0 & 0 & 0 & 0 & 0 & 0 \\
 & 1 & 0 & 0 & 0 & 0 & 0 & 1 & 0 & 0 & | & 0 & 0 & 0 & 0 & 0 & 0 & 0 & 0 \\
 & 1 & 0 & 0 & 0 & 0 & 0 & 0 & 1 & 0 & | & 0 & 0 & 0 & 0 & 0 & 0 & 0 & 0 \\
 & 0 & 1 & 0 & 0 & 1 & 0 & 0 & 0 & 0 & | & 0 & 0 & 0 & 0 & 0 & 0 & 0 & 0 \\
 & 0 & 1 & 0 & 0 & 0 & 1 & 0 & 0 & 0 & | & 0 & 0 & 0 & 0 & 0 & 0 & 0 & 0 \\
 & 0 & 1 & 0 & 0 & 0 & 0 & 1 & 0 & 0 & | & 0 & 0 & 0 & 0 & 0 & 0 & 0 & 0 \\
 & 0 & 1 & 0 & 0 & 0 & 0 & 0 & 1 & 0 & | & 0 & 0 & 0 & 0 & 0 & 0 & 0 & 0 \\
 & 0 & 0 & 1 & 0 & 1 & 0 & 0 & 0 & 0 & | & 0 & 0 & 0 & 0 & 0 & 0 & 0 & 0 \\
 & 0 & 0 & 1 & 0 & 0 & 1 & 0 & 0 & 0 & | & 0 & 0 & 0 & 0 & 0 & 0 & 0 & 0 \\
 & 0 & 0 & 1 & 0 & 0 & 0 & 1 & 0 & 0 & | & 0 & 0 & 0 & 0 & 0 & 0 & 0 & 0 \\
 & 0 & 0 & 1 & 0 & 0 & 0 & 0 & 1 & 0 & | & 0 & 0 & 0 & 0 & 0 & 0 & 0 & 0 \\
 & 0 & 0 & 0 & 1 & 1 & 0 & 0 & 0 & 0 & | & 0 & 0 & 0 & 0 & 0 & 0 & 0 & 0 \\
 & 0 & 0 & 0 & 1 & 0 & 1 & 0 & 0 & 0 & | & 0 & 0 & 0 & 0 & 0 & 0 & 0 & 0 \\
 & 0 & 0 & 0 & 1 & 0 & 0 & 1 & 0 & 0 & | & 0 & 0 & 0 & 0 & 0 & 0 & 0 & 0 \\
 & 0 & 0 & 0 & 1 & 0 & 0 & 0 & 1 & 0 & | & 0 & 0 & 0 & 0 & 0 & 0 & 0 & 0
\end{bmatrix}$$

is obtained. As can be seen, $\overline{A}_6$ only has entries different from "0" in the block which corresponds to the components of the tree. Each row now represents a minimal cut set, and the matrix contains all of them (compare with Section VI.B). It should be mentioned that the procedure just outlined gives all the cut sets of the tree. These do not necessarily have to be minimal, as in the present example. For this reason in general nonminimal cut sets have to be eliminated after the cut sets have been obtained. This applies as well if a minimal cut set appears several times. Then it is retained only once. Both elimination steps are realized using Boolean operations.

If a fault tree has a great number of cut sets, the foregoing method may require the application of a truncation procedure in order not to use excessive memory space and

computing time. This procedure may be qualitative, i.e., cut sets with more than a predetermined number of elements are not taken into account, or quantitative, i.e., cut sets with
an unreliability below a specified limit are suppressed. For details see Reference 39.

F. Comparison of the Methods

The direct simulation of reliability parameters is a flexible procedure for treating complex
systems, which takes into account, for example, complicated maintenance strategies, re-
strictions of repair capacity, the activation of reserve systems, etc. with ease. On the other
hand, the result is only obtained within confidence bounds, which in the case of very reliable
systems can only be narrowed at the expense of a great number of trials. The possibilities
of remedying this situation by applying variance reducing methods are rather limited.

Methods based on obtaining minimal cut sets permit a deeper insight into the system
structure. Analytical methods, in addition, yield exact solutions or permit, if simplifications
are required, a quantification of the error. On the other hand, sophisticated maintenance
strategies can either not be treated or require cumbersome mathematics. Fault trees with a
great number of cut sets may lead to problems with memory capacity and computing time.
For these reasons it is recommended that one have algorithms of both types available for
practical work, since no general rule exists as to which procedure is best suited for a specific
type of fault tree.

G. Calculation of the Probability and Expected Frequency of the Undesired Event from Minimal Cut Sets

The probability of occurrence of the undesired event is calculated by forming the expectation of the structure function given in Equation 14. Before doing this, one must observe
that the powers of binary variables, which appear in the products, are equal to the binary
variables themselves, i.e.,

$$x_i^m = x_i \ (m \neq 0) \tag{18}$$

and that all the powers must be replaced by the variable as such. For this reason, forming
the expectation of the structure function is equivalent to replacing the binary variables by
their probabilities, which may be constant or calculated, according to Equations 2, 3, or 5,
for example, depending on whether maintenance or inspection is included or not. This may
be done because the expectation of the product of independent random variables is equal to
the product of their expectations and that of their sum equal to the sum of the expectations.
Usually the expectation of Equation 14 is calculated after multiplying out its terms, which
gives:

$$\psi(\vec{x}) = \sum_{l=1}^{L} \kappa_l - \sum_{l=1}^{L-1} \sum_{m=l+1}^{L} \kappa_l \cdot \kappa_m + \sum_{l=1}^{L-2} \sum_{m=l+1}^{L-1} \sum_{n=m+1}^{L} \kappa_l \cdot \kappa_m \cdot \kappa_n$$

$$+ \ldots + (-1)^{L-1} \cdot \kappa_1 \cdot ,\ldots, \cdot \kappa_L \tag{19}$$

If in Equation 19 the expectation of the first term on the right side is calculated, an upper
bound for the occurrence probability is obtained. A lower bound results if the first two are
considered and so on.[24] If the probabilities to be used are small, the first term alone furnishes
a satisfactory approximation to the true result. If this is not the case, a closer upper bound
should be calculated as shown in Reference 38.

If initiating events are taken into account, Equation 19 is evaluated for each of them
separately using unavailabilities for the basic events. This is conveniently done by setting
the probability of occurrence of all initiating events to 0 with the exception of the one for
which the evaluation takes place. Its probability is set equal to 1. The result then obtained

is the unavailability of the system with respect to the initiating event in question. The corresponding expected frequency for occurrence of the undesired event as a consequence of the initiating event is obtained by multiplying this unavailability with the frequency, with which the initiating event is expected to happen. This may be given by its failure rate λ.

Uncertainties of reliability data may be propagated through the fault tree by performing a Monte Carlo calculation. Uniformly distributed random numbers are used to calculate a set of failure probabilities and failure rates on the basis of the distribution described in Section V.E. In doing this, the following relation may be used to calculate a failure rate (the procedure for an unavailability is analogous).

$$\lambda_i = \lambda_{50,i} \cdot \exp[\sqrt{-2 \cdot \ln z_p} \cdot \cos(2 \cdot \pi \cdot v_p) \cdot s_i] \tag{20}$$

In Equation 20, z_p and v_p are random numbers uniformly distributed in the interval 0,1. The remaining parameters are explained in Section V.E. The values resulting from Equation 20 are then used to evaluate Equation 19. Several trials are carried out from which the mean value and the standard deviation of the probability of occurrence of the undesired event are calculated, e.g., for the unreliability according to:

$$\bar{q}_s = \frac{1}{p} \cdot \sum_{p=1}^{P} q_{s,p} \tag{21}$$

with the standard deviation:

$$\sigma \bar{q}_s = [(\frac{1}{p} \cdot \sum_{p=1}^{P} q_{s,p}^2 - \bar{q}_s^2)/P]^{1/2} \tag{22}$$

In Equations 21 and 22, P is the total number of trials and $q_{s,p}$ the unreliability of the system calculated in trial p. Since the trials furnish the probability density function of the result, centiles may be obtained as well.

It should be noted that the foregoing treatment of uncertainties only reflects uncertainties owing to input data. No procedure exists to take into account uncertainties which may be originated by the modeling process, e.g., lack of completeness of the model.

VII. CASE STUDY

A. Introduction

Based on the study of Reference 40 the fault tree analysis of the process step nitration of a plant for fabricating hexogen is presented leaving out those parts which have not contributed significantly to the undesired event, an explosion of the reactor. In the analysis only the continuous operation and not the weekly startups and shutdowns are considered. The piping and instrumentation diagram of the process is shown in Figure 6.

Hexogen is produced according to the SH process[41,42] in which hexamine is nitrated with an excess of about 8 to 10 of highly concentrated nitric acid (98.5%). Reaction temperature should not exceed 20°C, otherwise a runaway reaction might occur. Apart from the reactants and the product ammonia, formaldehyde and other substances are present in the reactor, as discussed in detail in Reference 41. The reaction is exothermic and the mixture is chemically unstable, so that reaction temperature and nitric acid excess should be maintained within the permissible range in order to avoid an explosion.

B. Short Description of the Process

Nitric acid is supplied to the reactor at a temperature of 5°C. Hexamine is introduced into

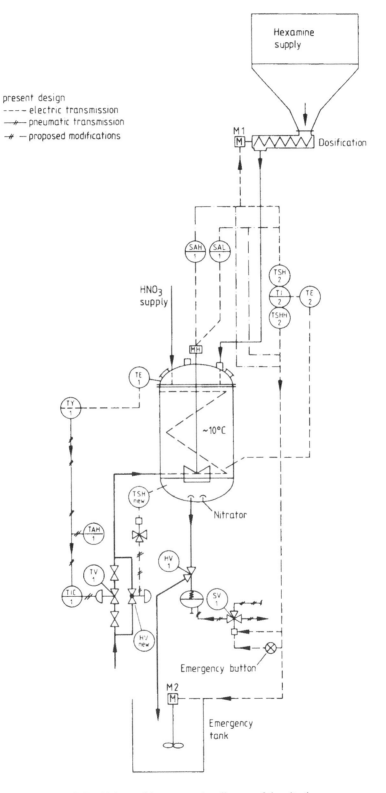

FIGURE 6.   Piping and instrumentation diagram of the nitration process.

the process by the dosification screw driven by the electric motor M1. The required ratio between the two is fixed manually at the start of operation. Because the reaction is exothermic the reactor requires cooling. This is provided by a mixture of water and methanol which passes through a coil inside the reactor with an inlet temperature of 5°C. The coolant mass flow is controlled by the pneumatic valve TV1 in order to maintain the reaction temperature at about 10°C, which is sufficient to allow the reaction to take place and far enough from the critical temperature. The position of valve TV1 is adjusted by a temperature control circuit consisting of the resistance thermometer TE1, the transducer TY1, which converts the electric signal from the thermometer into a pneumatic one, and the temperature indicator and controller TIC1, which provides the necessary position signal for valve TV1. In order to obtain a mixture as homogenous as possible and to avoid local heating, the reactor is equipped with a stirrer whose speed can be varied within a certain range.

C. Safety Equipment

From the process description it is clear that the following conditions have to be satisified in order to keep the hazard potential of the substances inside the reactor under control: (1) the excess of nitric acid has to be guaranteed, (2) nitration temperature has to be maintained close to 10°C, and (3) local heating has to be avoided by stirring. The essential safety features are

Reaction temperature: The temperature control circuit is equipped with the temperature alarm TA1 whose signal should prompt the operator to stop the hexamine supply and discharge the reactor to the emergency tank, which is filled with water. On discharge the tank stirrer driven by the motor M2 is started by pushing a button.

Stirrer revolutions: If the number of revolutions of the stirrer falls below the lower limit, the instrument SAL1 sets off an alarm in the control room. Should the number of revolutions rise above an upper limit, for example, because the stirrer shaft has broken, an alarm is activated by the instrument SAH1.

Discharge to the emergency tank: Activation of the discharge to the emergency tank is prompted by an electric measuring chain which is independent from operational procedures and powered by batteries which are constantly charged from the grid, so that power is available even if the grid failed. This chain consists of resistance thermometer TE2, temperature switch TSH2, which stops the hexamine supply if its limiting temperature is exceeded, and temperature switch TSHH2, which activates the discharge. In addition, the instrument TI2 indicates the reactor temperature in the control room. For discharging, solenoid valve SV1 is opened, which in turn opens the discharge valve HV1. At the same time, the stirrer in the emergency tank is turned on.

In addition to the automatic discharge just described, the reactor may be discharged by pushing a button, placed in both the control room and the plant, or by opening the valve HV1 with a handwheel.

D. Fault Tree Analysis

1. Development of the Fault Tree and the Reliability Data Base

The analysis performed with the program described in Sections III.D and E leads to the initiating events, which are presented in Table 1 along with their expected frequencies. The fault tree for the system is shown in Figure 7. It differs from that of the original work in that events 7 (coolant is not supplied or not supplied in adequate conditions) and 12 (HNO_3 supply below permissible limit) are not developed in detail because, as the analysis showed, they make a very small contribution to the expected frequency of an explosion. In the case of HNO_3 supply this is due to a highly redundant control and for the coolant supply to buffer deposits which provide an ample margin of time for possible countermeasures. As can be

Table 1
INITIATING EVENTS AND THEIR
EXPECTED FREQUENCIES OF
OCCURRENCE

Initiating event no.	Description	Expected frequency of the initiating event s_j in a^{-1}
3	Mechanical failure of control valve TV1	$2.5 \cdot 10^{-1}$
4	Failure of controller TIC1	$3.8 \cdot 10^{-1}$
7	No coolant supply or inadequate supply	$6.3 \cdot 10^{-2}$
10	Temperature measurement TE1 fails	$3.1 \cdot 10^{-2}$
11	Transducer TY1 fails	$5.5 \cdot 10^{-1}$
12	HNO₃ supply below permissible value	$5.0 \cdot 10^{-7}$
13	Stirrer shaft rupture	$1.8 \cdot 10^{-3}$
16	Failure of hydraulic stirrer motor	$8.8 \cdot 10^{-3}$
17	Failure of hydraulic supply	$7.1 \cdot 10^{-2}$
18	Coolant ingress into reactor	$8.5 \cdot 10^{-4}$

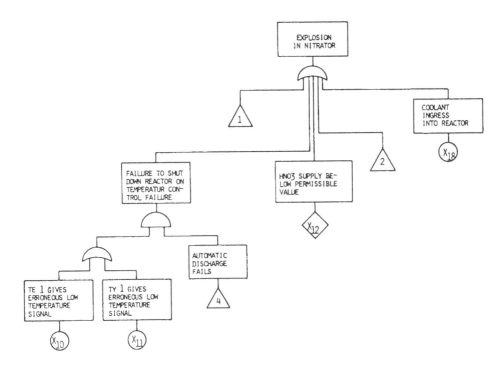

FIGURE 7.   (A to F) Fault tree for an explosion in the nitrator.

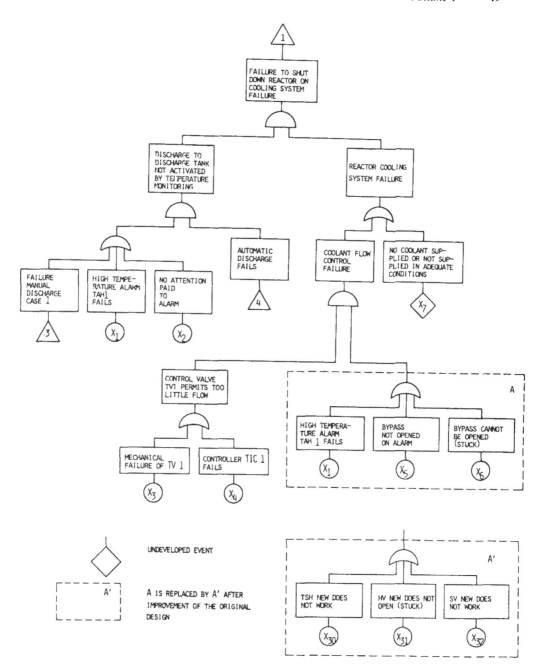

FIGURE 7B.

seen from Figure 7 the remaining events refer to disturbances in coolant mass flow control and to failures in stirring.

Since sources on reliability data for process plants are scarce, values from References 28 and 29 were combined with results from evaluations in nuclear power stations from Reference 22 to prepare a data base for the analysis. Whenever media other than water had to be considered and only values for components exposed to water were available, the possible increase of corrosion and wear was taken into account by multiplying the failure rates with environmental factors between 2 and 4. The data thus obtained are given in Table 2. Table

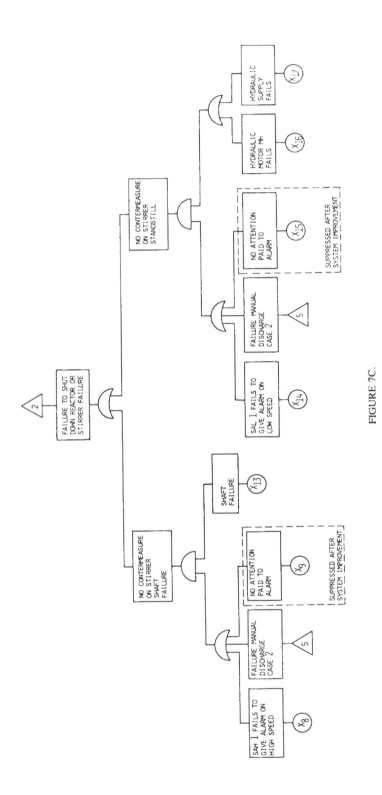

FIGURE 7C.

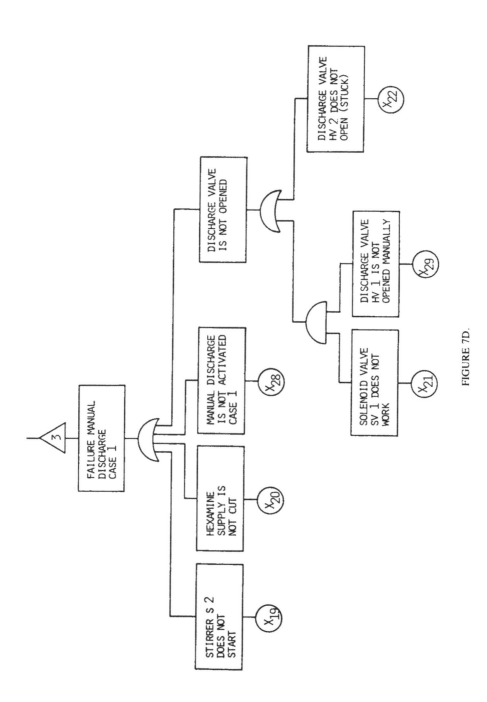

FIGURE 7D.

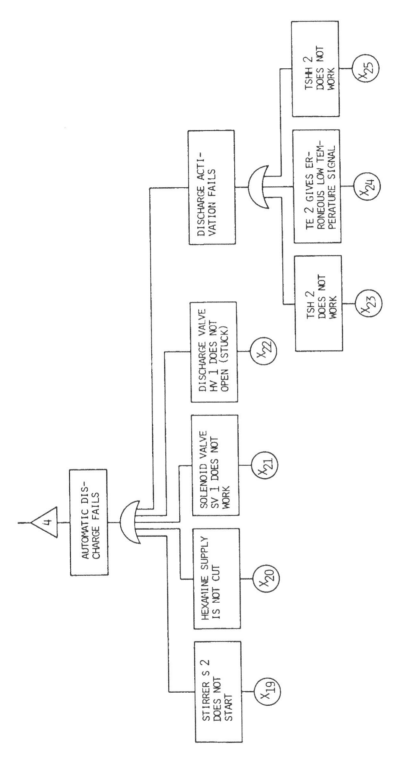

FIGURE 7E.

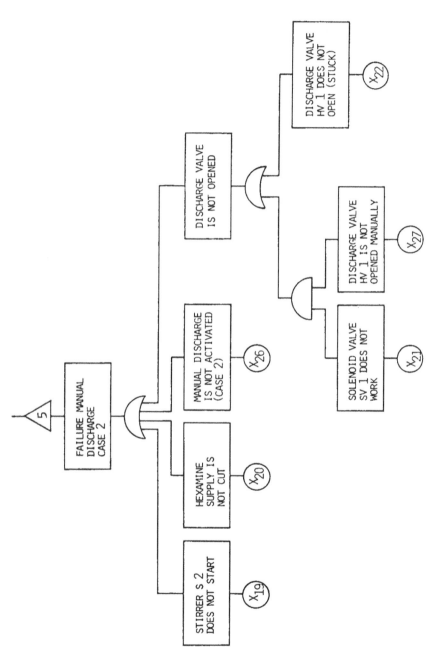

FIGURE 7F.

Table 2
FAILURE RATES FOR TECHNICAL COMPONENTS

Event no.	Description	Median in 10^{-6}/hr	Mean in 10^{-6}/hr	Factor of dispersion
1	Temperature alarm TAH1 fails low[a]	6.5	9.5	4.2
3	Mechanical failure of control valve TV1 producing too little flow[b]	21	29	3.6
4	Failure of controller TIC1 producing too little flow[b]	43	44	1.5
6	Bypass cannot be opened (stuck)	0.2	0.25	3.2
7	No coolant supply or supply not in adequate conditions (undeveloped event)[b]	7.2	7.2	1
8	Alarms for too many revolutions SAH1 fails[a]	7.9	13	5.0
10	Temperature measurement TE1 fails giving too low a temperature[b]	1.9	3.5	6.0
11	Transducer TY1 fails giving an output signal below actual temperature[b]	58	63	2.0
12	HNO$_3$ supply below permissible value (undeveloped event)[b]	$6 \cdot 10^{-5}$	$6 \cdot 10^{-5}$	1.0
13	Stirrer shaft breaks[b]	0.08	0.2	10
14	Alarm for too few revolutions SAL1 fails[a]	7.9	13	5
16	Failure of hydraulic stirrer motor[b]	0.38	1.0	10
17	Failure of hydraulic supply[b]	3.0	8.0	10
18	Coolant ingress into reactor (coil rupture)[b]	0.06	0.1	5.0
19	Stirrer motor M2 does not start	4.3	5.4	3.0
20	Hexamine supply is not cut Motor does not stop[c]	1.4	2.0	4.0
21	Solenoid valve SV1 does not work[c]	21	32	4.5
22	Discharge valve HV1 does not open[c]	1.7	3.0	6.0
23	Temperature switch TSH2 does not work[a]	8.4	12	4.0
24	Resistance thermometer TE2 gives too low a temperature	1.9	3.5	6.0
25	Temperature switch TSHH2 does not work[a]	8.4	12	4.0

Note: Usual time between inspections, θ = 672 hr.

[a]   A calibration error with a median u = 0.01 and a dispersion factor K = 5 is added in the calculations.
[b]   Initiating events.
[c] θ = 168 hr.

3 contains values for human error which were estimated according to the procedure described in Section V.D. In particular, the following arguments were used to fix the values:

• Basic event no. 2: no attention paid to alarm TAH1. For this simple act of recognition a failure probability with a median of $5 \cdot 10^{-4}$ and a dispersion factor of 10 is considered adequate.

• Basic event no. 5: bypass B1 is not opened on alarm. The operator disposes of about 10 min for diagnosing the disturbance after receiving the alarm signal. Danger increases during this period. The opening of the bypass valve has to be carried out close to the reactor at a high level of danger. For these reasons a failure probability with a median

Table 3
PROBABILITIES FOR HUMAN ERROR

Event no.	Description	Median in 10^{-6}	Mean in 10^{-6}	Factor of dispersion
2	No attention paid to alarm TAH1	500	1,332	10
5	Bypass B1 is not opened on alarm	500,000	806,967	5.0
9	No attention paid to alarm SAH1	500	1,332	10
15	No attention paid to alarm SAL1	500	1,332	10
26	Discharge not activated manually (case 2)	50,000	80,697	5.0
27	Discharge valve HV1 is not opened manually (case 2)	1 million	1 million	1.0
28	Discharge not activated manually (case 1)	5,000	1,332	10
29	Discharge valve HV1 is not opened manually (case 1)	500,000	806,967	5.0

of 0.5 and a dispersion factor of 5 is assumed (the lognormal distribution has to be cut off at the maximum possible value of a probability, i.e., 1).

- Basic events no. 9 and no. 15: no attention is paid to alarms SAH1 or SAL1. Since this is only an act of recognition a failure probability of $5 \cdot 10^{-4}$ with a dispersion factor of 10 is estimated.
- Basic event no. 26: discharge is not activated manually (case 2). Only 5 min are available for this intervention which is required when an alarm for high or low stirrer speed or high temperature in the reactor sounds. The discharge may be activated either from the control room or the plant itself and takes place under high stress. Therefore, a failure probability of $5 \cdot 10^{-2}$ with a dispersion factor of 5 is assumed.
- Basic event no. 27: discharge valve HV2 is not opened manually (case 2). This intervention consists in turning a handwheel. It has to be carried out close to the reactor under extreme danger. Therefore, it is believed that it will not be realized at all and a failure probability of 1 with a dispersion factor of 1 is chosen.
- Basic event no. 28: discharge is not activated manually (case 1). This measure differs from that of event no. 26 in that 1 hr is available for its execution. Since it is a safety measure, it is assumed that the operator is well acquainted with it. Therefore, a failure probability of $5 \cdot 10^{-3}$ with a dispersion factor of 10 is chosen.
- Basic event no. 29: discharge valve HV2 is not opened manually (case 1). The action is the same as required in event no. 27. The situation however, is less dangerous. Therefore, a probability of 0.5 with a dispersion factor of 5 is considered adequate (the lognormal distribution is cut off at the maximum possible value of a probability, i.e., 1).

2. Results of the Analysis

With the reliability data from Tables 2 and 3 and the expected initiating event frequencies indicated in Table 1 the results for the expected frequencies of an explosion in the nitrator are obtained, as shown in Table 4. The total frequency of the event is

$$h = 4 \cdot 10^{-2} \, a^{-1}$$

The major contributions to this value are made by the following initiating events:

- no. 11: Transducer TY1 fails (62.5%)
- no. 17: Failure of hydraulic supply (19%)

Table 4
EXPECTED FREQUENCIES OF AN EXPLOSION IN THE NITRATOR
(VALUES FOR INDIVIDUAL INITIATING EVENTS AND TOTAL VALUE)

Initiating event no.	Description	Expected frequency of the initiating event s_j in a^{-1}	Unavailability u_j	Expected frequency of the undesired event h_j in a^{-1}
3	Mechanical failure of control valve TV1	$2.5 \cdot 10^{-1}$	$4.8 \cdot 10^{-3}$	$1.2 \cdot 10^{-3}$
4	Failure of controller TIC1	$3.8 \cdot 10^{-1}$	$4.8 \cdot 10^{-3}$	$1.8 \cdot 10^{-3}$
7	No coolant supply or inadequate supply	$6.3 \cdot 10^{-2}$	$5.7 \cdot 10^{-3}$	$3.6 \cdot 10^{-4}$
10	Temperature measurement TE1 fails	$3.1 \cdot 10^{-2}$	$4.6 \cdot 10^{-2}$	$1.4 \cdot 10^{-3}$
11	Transducer TY1 fails	$5.5 \cdot 10^{-1}$	$4.6 \cdot 10^{-2}$	$2.5 \cdot 10^{-2}$
12	HNO_3 supply below permissible value	$5.0 \cdot 10^{-7}$	1.0	$5.0 \cdot 10^{-7}$
13	Stirrer shaft rupture	$1.8 \cdot 10^{-3}$	0.11	$2.0 \cdot 10^{-4}$
16	Failure of hydraulic stirrer motor	$8.8 \cdot 10^{-3}$	0.11	$9.7 \cdot 10^{-4}$
17	Failure of hydraulic supply	$7.1 \cdot 10^{-2}$	0.11	$7.8 \cdot 10^{-3}$
18	Coolant ingress into reactor	$8.5 \cdot 10^{-4}$	1.0	$8.5 \cdot 10^{-4}$
Total				$4.0 \cdot 10^{-2}$

- no. 4: Failure of controller TIC1 (5%)
- no. 10: Failure of resistance thermometer TE1 (4%)
- no. 3: Mechanical failure of control valve TV1 (3.3%)
- no. 16: Failure of hydraulic stirrer motor (2.4%)

It is clear from Table 4 that an initiating event makes a small contribution to the expected frequency of an explosion, if it occurs seldom, if the operating and safety systems available to cope with it have a low unavailability, or if both are true. Relatively high unavailabilities are encountered in the case of events 10, 11, 13, 16, and 17. For this reason the corresponding minimal cut sets are presented in Table 5. For both groups of events there are a number of minimal cut sets which contain just one additional event to the intiating event. Especially high contributions are made in the case of initiating events 10 and 11 by the failure of the temperature switches of the emergency discharge system, TSH2 and TSHH2, and for initiating events 13, 16, and 17 by the operator intervention represented by basic event no. 26 and the failure of the speed alarms SAH1 and SAL1.

For these reasons the following proposals for system improvement are made: installation of an independent temperature switch for opening the bypass of the cooling system in case of high temperatures in the reactor. In this way a system which is redundant to the operational system would be created. In addition, the necessity of discharging the reactor contents into the emergency tank in case of disturbances in the coolant control system would frequently not be given and the product would be saved. The modification can easily be put into practice. It is indicated in Figure 6 and its effect on the fault tree is shown in Figure 7.

Table 5
MINIMAL CUT SETS WITH A HIGH UNAVAILABILITY

In addition to the initiating events the minimal cut sets contain the basic event(s)	Time averaged unavailability
Initiating Events 10 or 11	
23	$2.0 \cdot 10^{-2}$
25	$2.0 \cdot 10^{-2}$
21	$2.7 \cdot 10^{-3}$
24	$1.2 \cdot 10^{-3}$
20	$1.7 \cdot 10^{-4}$
22	$2.6 \cdot 10^{-4}$
19	$1.8 \cdot 10^{-3}$
Total	$4.6 \cdot 10^{-2}$
Initiating Events 13, 16, or 17	
26	$8.1 \cdot 10^{-2}$
14	$2.0 \cdot 10^{-2}$
21,27	$2.7 \cdot 10^{-3}$
19	$1.8 \cdot 10^{-3}$
15	$1.3 \cdot 10^{-3}$
Total	0.11

Note: This applies to initiating events 16 and 17. In the case of initiating event 13, basic event 15 has to be replaced by basic event 9 with the same numerical value.

In order to quantify the new configuration, the following additional reliability data were used:

Temperature switch: $\lambda_{50} = 3.9 \cdot 10^{-6}/\mathrm{hr}, \ K = 6$

Inspection interval $\theta = 672$ hr

Solenoid valve: $\lambda_{50} = 21 \cdot 10^{-6}/\mathrm{hr}, \ K = 4.5$

Inspection interval $\theta = 168$ hr

Control valve: $\lambda_{50} = 29 \cdot 10^{-6}/\mathrm{hr}, \ K = 5$

Inspection interval $\theta = 168$ hr

In addition, a possible calibration error of the temperature switch is taken into account with a probability of $u_{50} = 0.01$ and a dispersion factor of 5.

The system modification has the following quantitiative effects: (1) the unavailability of the systems required to cope with initiating events 10 and 11 drops from $4.6 \cdot 10^{-2}$ to $1.2 \cdot 10^{-3}$; (2) the unavailability of the systems required to cope with initiating events 3 and 4 is reduced from $5 \cdot 10^{-3}$ to $1.6 \cdot 10^{-4}$.

In the case of disturbances related to the stirring of the reactor the most important contribution stems from the manual activation of the reactor discharge (basic event no. 26). If the discharge is activated automatically by alarms SAL1 and SAH1, the unavailability of the systems required to cope with the corresponding initiating events drops from 0.11 to $2.5 \cdot 10^{-2}$. Again, the modification can be realized at little expense.

The modifications which have been proposed reduce the expected frequency of an explosion in the reactor from

$$h = 4.0 \cdot 10^{-2} \, a^{-1} \quad \text{to} \quad h' = 4.1 \cdot 10^{-3} \, a^{-1}$$

In the new configuration the major contributions to the explosion frequency are made by disturbances related to the stirring of the reactor, viz., 51%. If uncertainties of input data are propagated as described in section VI, G., the following 5 and 95% centiles are obtained:

$$\text{Original design: } h_{o5} = 8.7 \cdot 10^{-3} \, a^{-1} \quad \text{and} \quad h_{95} = 0.1 \, a^{-1}$$

$$\text{Modified design: } h'_{o5} = 3.6 \cdot 10^{-4} \, a^{-1} \quad \text{and} \quad h'_{95} = 1.3 \cdot 10^{-2} \, a^{-1}$$

The improvement of the system resulting from the proposed modifications can be considered as real and not inside the range of statistical uncertainty, since mean values and centiles after improvement lie below the corresponding values of the original design.

E. Conclusions

The analysis has provided guidance for an improvement of system safety and, although this had not been the express aim of the investigation, of its availability, thus increasing its economic efficiency. Some of the results were already obtained in the qualitative phase of the analysis. The quantification of the fault tree has brought further insights and uncovered areas where safety measures were imbalanced (strongly different contributions to the expected frequency of an explosion from the different initiating events). The proposals for improving the plant lead to a reduction of the expected frequency of an explosion by a factor of 10 and can be realized at little expense. An interpretation of the results as absolute values is impeded by flaws still existent in reliability data for process plants. On the other hand, it should be noted that among the components of important systems such as temperature control and reactor discharge only the resistance thermometers, and in the case of the latter, the discharge valve as well, are subject to typically chemical exposure. For this reason, results obtained for these areas with reliability data from other industries may be considered as more trustworthy than those for other parts of the plant.

REFERENCES

1. **Powers, G. J. and Tompkins, F. C., Jr.,** Fault tree synthesis for chemical processes, *AIChE J.*, 20(2), 376, 1974.
2. **Menzies, R. M. and Strong, R.,** Some methods of loss prevention, *Chem. Eng.*, p. 151, March 1979.
3. **Taylor, J. R.,** *A Background to Risk Analysis*, Vol. 1-4, Risø Research Centre, Roskilde, Denmark, 1979.
4. **Hope, S., Bjordal, E. N., Diack, H. M., Eddershaw, B. W., Joanny, L., Ortone, G., Payne, F. G., Searson, A. H., Sedlacek, K. W., and van Strien, W.,** Methodologies for Hazard Analysis and Risk Assessment in the Petroleum Refining and Storage Industry, CONCAWE, Den Haag, The Netherlands, 1982.
5. **Vesely, W. E., Goldberg, F. F., Roberts, N. H., and Haasl, D. F.,** Fault Tree Handbook, U.S. Nuclear Regulatory Commission, Washington, D.C., 1981.

6. **Hauptmanns, U.**, Análisis de Arboles de Fallos, editorial bellaterra, Barcelona, 1986.
7. PRA Procedures Guide — A Guide to the Performance of Probabilistic Risk Assessments for Nuclear Power Plants, NUREG/CR-2300, Vol. 1 and 2, U.S. Nuclear Regulatory Commission, Washington, D.C., 1983.
8. U.S. Nuclear Regulatory Commission, Reactor Safety Study — An Assessment of Accident Risks in U.S. Commercial Power plants, WASH-1400, NRC, Washington, D.C., 1975.
9. Der Bundesminister für Forschung und Technologie, Deutsche Risikostudie Kernkraftwerke-Eine Untersuchung zu dem durch Störfalle in Kernkraftwerken verursachten Risiko, Gesellschaft für Reaktorsicherheit, Cologne, 1979.
10. **Kumaomoto, H. and Henley, E. J.**, Protective system hazard analysis, *Ind. Eng. Chem. Fundam.*, 17(4), 274, 1978.
11. **Nielsen, D. S., Platz, O. and Kongsø, H. E.**, Reliability Analysis of a Proposed Instrument Air System, RISØ-M-1903, Risø Research Centre, Roskilde, Denmark, 1977.
12. **Shaeiwitz, J. A., Lapp, S. A., and Powers, G. J.**, Fault tree analysis of sequential systems, *Ind. Eng. Chem. Process Des. Dev.*, 16(4), 529, 1977.
13. **Hauptmanns, U.**, Fault tree analysis of a proposed ethylene vaporization unit, *I & EC Fundam.*, 19(3), 300, 1980.
14. **Hauptmanns, U., Yllera, J., and Sastre, H.**, Safety analysis for the ammonia-air mixing system of a plant for the production of nitric acid, *J. Chem. Eng. Jpn.*, 15(4), 286, 1982.
15. **Hauptmanns, U. and Sastre, H.**, Safety analysis of a plant for the production of vinyl acetate, *J. Chem. Eng. Jpn.*, 17(2), 165, 1984.
16. Risk Analysis of Six Potentially Hazardous Industrial Objects in the Rijnmond Area, A Pilot Study, D. Reidel, Dordrecht, The Netherlands, 1982.
17. **Caldarola, L.**, Fault Tree Analysis with Multistate Components, KfK 2761, EUR 5756e, Karlsruhe, West Germany, 1979.
18. **Caldarola, L.**, Generalized Fault Tree Analysis Combined with State Analysis, KfK 2530, EUR 5754e, Karlsruhe, West Germany, 1980.
19. **Hayes-Roth, F., Waterman, D. A., and Lenat, D. B., Eds.**, *Building Expert Systems*, Addison-Wesley, London, 1983.
20. **Sax, N. I.**, *Dangerous Properties of Industrial Materials*, van Norstrand Reinhold, New York, 1984.
21. **Roth, L. and Weller, U.**, Gefährliche Chemische Reaktionen, Ecomed Verlagsgesellschaft mbH, Landsberg, West Germany, 1982.
22. **Hömke, P., Krause, H. W., Ropers, W., Verstegen, C., Schlenker, H. V., Hüren, H., Dörre, D., and Tsekouras, A.**, Zuverlässigkeitskenngrößenermittlung im Kernkraftwerk Biblis — Abschlußbericht, Rep. no. GRS-A-1030, Gesellschaft fur Reaktorsicherheit, Cologne, 1984.
23. **Kletz, T. A.**, Plant instruments: which ones don't work and why, *Chem. Eng. Progr.*, 76(7), 68, 1980.
24. **Barlow, R. E. and Proschan, F.**, *Statistical Theory of Reliability and Life Testing — Probability Models*, Holt, Rinehart, & Winston, New York, 1975.
25. **Gaede, K. W.**, *Zuverlässigkeit — Mathematische Modelle*, Carl Hanser Verlag, Munich, 1977.
26. **Cox, D. R.**, *Renewal Theory*, Methuen, London, 1962.
27. **Schneeweiss, W.**, *Zuverlässigkeitstheorie*, Springer-Verlag, Berlin, 1973.
28. **Anyakora, S. N., Engel, G. F. H., and Lees, F. P.**, Some data on the reliability of instruments in the chemical plant environment, *Chem. Eng.*, p. 396, November 1971.
29. **Skala, V.**, Improving instrument service factors, *Instrum. Technol.*, p. 27, November 1974.
30. **Fleming, K. N. and Raabe, P. H.**, Comparison of three methods for the quantitative analysis of common cause failure, Proc. ANS Topical Meet. Probabilistic Analysis of Nuclear Reactor Safety, GA-A-14568, General Atomic Co., San Diego, Calif. 1978.
31. **Vesely, W. E.**, Estimating common cause failure probabilities in reliability and risk analyses: Marshall-Olkin specializations, in *Nuclear Systems Reliability and Risk Assessment*, Fussel, J. B. and Burdick, G. R., Eds., Society for Industrual and Applied Mathematics, Philadelphia, 1977.
32. **Swain, A. D. and Guttman, H. E.**, Handbook on Human Reliability Analysis with Emphasis on Nuclear Power Plant Applications, NUREG/CR-1278, U.S. Nuclear Regulatory Commission, Washington, D.C., 1983.
33. **Hammersley, J. M. and Handscomb, D. C.**, *Monte Carlo Methods*, Methuen, London, 1975.
34. **Hauptmanns, U. and Yllera, J.**, Fault-tree evaluation by Monte Carlo simulation, *Chem. Eng.*, p. 91, January 1983.
35. **Kamarinopoulos, L.**, Direkte und Gewichtete Simulationsmethoden zur Zuverlässigkeitsuntersuchung Technischer Systeme, Doctoral thesis, Technical University of Berlin, 1972.
36. **Kamarinopoulos, L.**, Anwendung von Monte-Carlo-Verfahren zur Ermittlung von Zuverlässigkeitsmerkmalen technischer Systeme, ILR-Bericht, Technical University of Berlin, 1976, 14.

37. **Güldner, W., Polke, H., Spindler, H., and Zipf, G.,** Computer Code Package RALLY for the Probabilistic Safety Assessment of Large Technical Systems, GRS-57, Gesellschaft für Reaktorsicherheit, Cologne, 1984.
38. **Hauptmanns, U.,** Métodos para la evaluación de árboles de fallos, *Energ. Nucl.,* (Spain), 23(122), 393, 1979.
39. **Camarinopoulos, L. and Yllera, J.,** An improved top-down algorithm combined with modularization as a highly efficient method for fault tree analysis, *Reliability Eng.,* 11, 93, 1985.
40. **Hauptmanns, U., Hömke, P., Huber, J., Reichart, G., and Riotte, H.-G.,** Ermittlung der Kriterien für die Anwendung Systemanalytischer Methoden zur Durchführung von Sicherheitsanalysen für Chemieanlagen, im Auftrag des Umweltbundesamtes, Berlin, GRS-59, Gesellschaft für Reaktorsicherheit, Cologne, West Germany, 1985.
41. **Urbanski, T.,** *Chemistry and technology of explosives,* Vol. 1-4, Pergamon Press, Oxford, 1964-1984.
42. **Lingens, P.,** Hexogen, in *Ullmann's Encyklopädie der Technischen Chemie,* Vol. 21, Bartholome, E., Biekert, E., Hellmann, H., Weigert, M., and Weise, E., Eds., Verlag Chemie, Weinheim, West Germany, 1982.

Chapter 4

HAZARD IDENTIFICATION AND SAFETY ASSESSMENT OF HUMAN-ROBOT SYSTEMS

Hiromitsu Kumamoto, Yoshinobu Sato, and Koichi Inoue

TABLE OF CONTENTS

I. INTRODUCTION

Industrial robots are releasing humans from hazardous, heavy, or monotonous labor. The robot itself, however, could become a source of hazard. At least four people are reported to have been killed by robots in Japan, which now installs more than 50% of the industrial robots of the world. Private information sources indicate that robots are suspected to have killed at least three other people. Injuries caused by robots are several times larger than the fatalities. Industrial safety is now being threatened by the robots that should have improved the welfare of the labor force.

In view of the dilemma, the Japanese government in June 1983 added to the OISH (Ordinance of Industrial Safety and Health) new rules and guidelines for safety designs and uses of robots in manufacturing industries.[1] The ordinance identified, as key elements, emergency shutdown devices and other protection features; however, there remain many open problems, including comprehensive safety studies of human-robot systems based on modern probabilistic risk assessment techniques being widely applied to nuclear and aerospace industries.

Currently, robots are mostly used in the manufacturing industries, the main target of the June 1983 ordinance. It is, however, predicted that in the very near future robots will infiltrate other industries such as construction, building, mining, agriculture, forestry, fishing, medicine, and service. Robots will surely cause as serious a nuisance as the automobile did in its early days.

The purpose of this chapter is to develop a systematic methodology for safety improvements of human-robot systems at present and for the future. Various types of robots are briefly summarized in Section II together with safety aspects specific to robots. Section III gives historical reviews of typical industrial incidents caused by robots. Seciton IV outlines safety assessment procedures which are consistent with the probabilistic risk assessment procedures typically used for systems in the nuclear, chemical, and aerospace industries. Section V introduces a character analysis and hazard causation mechanisms are analyzed for a general human-machine system. Section VI then considers the human-robot system, and enumerates possible hazards as a subset of the ones obtained in Section V. Two typical hazards, ie., human struck by robot body and human struck by robot arm, are targeted in Sections VII and VIII, and logic models as a general purpose fault tree are constructed. Section IX outlines phased mission analysis for installation, startup, test, instruction, operation, maintenance, etc., and generates minimal cut sets of the fault trees for typical phases. A realistic human-robot system is assessed quantitatively in Section X by a computer program based on a version of KITT (Kinetic Tree Theory).[2,3] This chapter is partially based on our papers published in the *Transactions of the Japanese Society of Mechanial Engineers*.[4,5]

II. ROBOT CLASSIFICATION AND SAFETY ASPECTS

There are a number of robot classification schemes from points of view of input, control, programmability, and motion. The Japanese Industrial Standard (JIS B 0134-1979) uses the following classifications:

1. Input: manual manipulator, fixed sequence robot, variable sequence robot, playback robot, numerically controlled robot, and intelligent robot.
2. Programmability: single-purpose robot, single-purpose programmed repeatable robot, and multipurpose programmed repeatable robot.
3. Motion: cylindrical coordinate robot, polar coordinate robot, Cartesian coordinate robot, and articulated robot.

The OISH[1] defines the industrial robot as "the machines (excluding those under research and development and those specified by the Ministry of Labor) which are equipped with manipulators and memory systems (including variable and fixed sequence control systems) and which allow manipulators to expand and contract, bend and stretch, move up and down, move from side to side, circle round, or perform a combination of these motions automatically according to information stored in memory systems..." The fixed sequence robot belongs to the industrial robot according to this definition, but the manual manipulator does not. Neither of them is classified as an industrial robot in the U.S. nor in Europe.

Safety aspects specific to the robots are

1. The arms and body can move relatively freely in the three-dimensional space often with a large power.
2. The automatic control feature makes the robot motion sometimes unpredictable from outside.
3. Tools, works, and materials become more hazardous when coupled with the robot movements.
4. Robot interactions with peripheral machines could trigger safety problems.
5. Instruction (teaching) or troubleshooting require humans to enter the robot movement zone while the main power switch is On. Similar intrusions could occur during the troubleshooting of peripheral machines around the robot.
6. Auxiliary tasks near the robot cannot be eliminated for supervision, tool exchange, test, etc.
7. Users exploit the robot in many ways. Consequently, it is not sufficient to rely solely on safety features provided by robot makers.
8. An automated manufacturing system sometimes consists of robots and machines made by different makers, and some characteristics as a system are difficult to predict.
9. Abnormal occurrence data have not been systematically collected, and one must rely on his/her narrower experience about robot safety.
10. Neither robot hardware nor software is completely reliable, and the two interact with each other.

III. HISTORICAL OVERVIEW OF ROBOT INCIDENTS

Fixed sequence robots were used in Japan during the early 1960s. Playback robots were then introduced from the U.S. to Japan in 1967, which triggered wider robot exploitation in industry today. There is a scarcity of historical records about robot incidents because it was not until recently that people began to recognize robots as a source of nuisance for human safety. Table 1 lists only a subset of robot incidents from the point of view of year, situation, cause, consequence, and operation phase.[1]

IV. BASIC STEPS FOR SAFETY ASSESSMENT

This chapter uses the term "safety assessment" in the following sense: a process consisting of (1) clarification of causal relations resulting in physical human hazards such as death, injury, or sickness; (2) determination of generation mechanisms and probabilities of these hazards; and (3) evaluation of original and modified systems with safety measures.

Two opposite approaches exist for the assessment of safety: one is inductive, and other deductive. The inductive approach can be applied to systems with long operation histories and thus enough experiences of system abnormalities. This statistically analyzes a large number of accidents, incidents, and other occurrences observed so far, and assesses the system safety accordingly. Highway traffic control systems and automobiles typically fit the statistical analysis and assessment.

Table 1
TYPICAL INDUSTRIAL ROBOT INCIDENTS

No.	Year	Situation and cause	Consequence	Phase
1	1978	A limit switch of a supply/removal robot was triggered when a workman intended to pick up a defective work. The robot pressed him at his back.	Fatality	Operation
2	1978	A troubleshooter entered the operation zone of a welding robot. He was struck by the robot manipulator in his stomach.	Injury	Operation
3	1979	A workman entered the operation zone of an assembly robot in order to pick up a part to be assembled. He did not turn off the main power switch of the robot, and the manipulator seized his fingers against peripheral equipment.	Minor	Operation
4	1979	A manipulator inadvertantly started moving during inspection of an assembly robot, and hit a workman on his hand.	Minor	Inspection
5	1981	A workman incorrectly pressed a start button after an alignment operation of a shaving machine. Consequently, a supply/removal robot behind him suddenly began to move and pressed him against the peripheral machine.	Fatality	Alignment
6	1981	A workman was aligning an assembly robot with its cover being open. His hand was caught by the transmission when he stretched the hand toward a component that had fallen into the robot.	Injury	Alignment
7	1981	A manipulator suddenly began to move during an instruction phase of a robot, and hit a workman.	Minor	Instruction
8	1981	A workman started cleaning up a peripheral machine without turning off the main power switch of a robot. He stretched his hands, being struck by the manipulator.	Minor	Operation
9	1981	A robot manipulator under repair inadvertently started moving and hit a repairman on his hand.	Minor	Repair
10	1981	A robot manipulator under repair moved and hit a repairman on his hand when a colleague erroneously turned on a switch.	Minor	Repair
11	1982	A workman with an air gun was removing waste shred from a work of a lathe. A work supply robot hit him on his hand.	Minor	Operation
12	1984	A workman incompletely switched a flexible transfer line into a manual mode. He entered a dangerous zone and a robot arm over his head severely pressed his head against a machining center.	Fatality	Operation

The deductive approach, which this chapter is concerned with, deals with relatively innovative systems where there is a scarcity of historical operation data. This approach first establishes causal models where less experienced events are successively expressed as functions of more experienced events, and ultimately as functions of basic or axiomatic events with enough statistical data. The human-robot systems fit the deductive framework because of innovative technologies and resultant new types of interactions among system components.

The deductive safety assessment consists of the following five fundamental steps.

1. Enumeration of hazards: This enumerates scenarios of events resulting in damages to humans.
2. Preliminary hazard evaluation: The hazards enumerated in Step 1 are prioritized and filtered according to loss magnitude, plausibility, and prevention possibility.
3. Synthesis and analysis of causal models: A deductive model typified by a fault tree is constructed with each hazard of importance being a top event. These models are analyzed qualitatively and/or quantitatively. The qualitative analysis yields system failure modes called minimal cut sets, clarifies how each hazard can occur, and identifies unsafe links in the system. The quantitative analysis typically calculates the expected number of occurrences of the hazard during a time period, and makes it possible for us to compare the number with familiar risk sources such as automobiles and lightning.
4. Evaluation of corrective measures: Corrective measures are introduced to mitigate the major hazards. The causal models in step 3 are modified accordingly, and the effectiveness of the measures are evaluated qualitatively and/or quantitatively based on reductions of loss magnitudes and occurrence probabilities. The corrective measures identified as acceptable are assessed in the next step.
5. System assessment: This is an extension of evaluation phases in steps 3 and 4, and comprehensive assessments are carried out with consideration of more factors. The system designed in step 4 is assessed by hazard types, occurrence probabilities, loss magnitudes, corrective measures effectiveness, costs, operability, maintainability, productivity, legal regulations, public attitudes, management philosophy, etc. Steps 4 and 5 are repeated until all the remaining hazards are assessed as acceptable. The system is regarded as too unsafe, and should be rejected if such a convergence cannot be attained.

V. HAZARD ENUMERATION

A. Introduction

This section develops a technique for hazard enumeration, the first step of the assessment of safety. The succeeding steps may be incorrect if one does not attend to hazards of importance at this step.

Hazard enumeration is a relatively undeveloped field with few systematic procedures compared with other steps of safety assessment. This is mainly due to:

1. Historically, the system safety assessment technique has been used mainly in the aerospace, nuclear power, and chemical industries where hazards occur in extreme forms such as airplane crashes, fission product release, plant explosion, etc. Compared with these hazards, events such as a person stumbling off toward a rotating machine have been neglected.
2. Most systems in these industries are single-purpose, yielding a smaller variety of hazards which can be intuitively identified without elaborate enumeration techniques.

The human-robot system, on the other hand, is a mobile computer-based configuration, and hence has a multipurpose flexibility, resulting in a large number of hazards according to the level of intelligence of the robot. This is a similar situation to the diversity of social crimes. A more comprehensive enumeration technique is required for the safety assessment of the human-robot system.

The hazard is defined in MORT (Management Oversight and Risk Tree)[6] as the potential in an activity (or condition or situation) for sequence(s) of errors, oversights, changes, and stresses to result in an unwanted transfer of energy with resultant damage to persons, objects, or processes. The "potential", "sequence(s)", "unwanted transfer of energy", and "resultant damage" are the key words of the definition.

This chapter is concerned with the safety problem of the human-robot system, and only the physical damage to persons such as death, injury, and sickness, excluding psychological annoyance or disorder is considered. The unwanted transfer of energy[6] is mostly a correct concept as the final causes of human damage, except for cases where energy or utility shortage, virus, bacteria, chemicals, or poisons are the cause. A more complete list of the final causes of human damage follows.

1.   Unwanted energy transfer: Kinetic, potential, heat, electric, or other energy transfer from a system component to human. For instance, the kinetic and potential energies cause injuries of various types. The heat energy yields burns. Electric shocks are caused by electricity, bradyacusia by sounds, and Raynaud's disease by vibrations. Various hazards are caused by radiation energies.

2.   Unwanted material transfer: Materials such as chemicals or microbes are transferred from a component to human.

3.   Unwanted energy blockage: Some transfers of energy to human are blocked inadvertently. People may die from cold temperatures.

4.   Unwanted material blockage: Materials such as air or water are blocked between a system component and the human.

The phrase "potential in an activity for sequence of" in the MORT definition is a little difficult to understand. This article replaces it by "scenarios". By phrasing the definition in a different way, we have a new definition: the hazard is a scenario about the system to result in an unwanted transfer or blockage of energy or material with resultant damage to persons. The errors, oversights, changes, stresses, transfers, and blockages are building blocks of the scenario. The hazard enumeration is now reduced to the enumeration of such scenarios. In the next section, we describe a heuristic technique which makes use of characters involved in the scenario.

B. Character Analysis

Let us consider a general human-machine system containing the human-robot system as a special case. In such a system, there is a sequence of events much like the "domino effect" when an accident occurs: a system component triggers an event, this triggers the next event, and so on. We can imagine three types of characters in the event sequence: victim, assailant, and accomplice. The victim, of course, is a human when system safety is a subject as in this chapter. The assailant is a system component ultimately causing damage to humans via energy transfer, material transfer, energy blockage, or material blockage. One or more accomplices are involved in complicated hazard scenarios. Also considered is a background on which the scenario is played. This typically involves performance shaping factors developed in THERP (Technique for Human Error Rate Prediction).[7]

In the scenario, a character has a certain correlation with another. Denote the correlation by a directed arrow which points from the first character to the second appearing on the stage. Number the arrows to denote the time series of the correlation. We can classify typical scenarios according to the number of correlations involved, as shown in Figure 1.

1.   Single-correlation scenario: Only ASS-VIC type exists where the assailant causes the damage to the victim.

2.   Double-correlation scenario: VIC-ASS-VIC — the victim first appears and is correlated with the assailant which, in turn, causes the damage to the victim. ACC-ASS-VIC — an accomplice other than the victim first shows up and is correlated with the assailant to cause the damage to the victim.

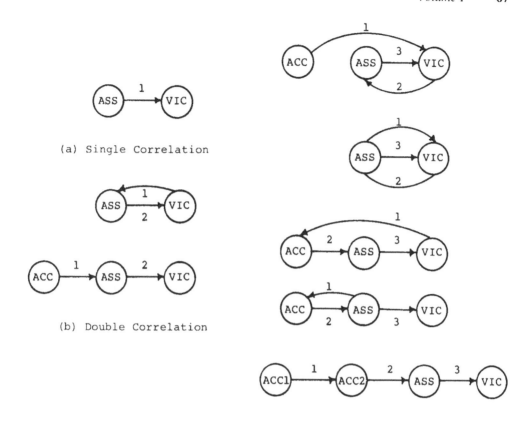

(a) Single Correlation

(b) Double Correlation

(c) Triple Correlation

FIGURE 1. Scenario classification by character analysis.

3. Triple-correlation scenario: ACC-VIC-ASS-VIC — the accomplice first shows up and triggers the double-correlation scenario VIC-ASS-VIC; in other words, the accomplice is first correlated with the victim which, in turn, triggers the assailant to cause the damage. ASS-VIC-ASS-VIC — the assailant triggers the double-correlation scenario VIC-ASS-VIC. VIC-ACC-ASS-VIC — the victim triggers the double-correlation scenario ACC-ASS-VIC. ASS-ACC-ASS-VIC — the assailant starts the double-correlation scenario ACC-ASS-VIC. ACC1-ACC2-ASS-VIC — another accomplice triggers the double-correlation scenario.

The classification can be extended recursively to four or more correlations by considering a character which appears first and triggers the scenarios with one less correlation.

The final correlation from the assailant to the victim is called ultimate. This is caused by energy/material transfer or blockage. The other correlations are called intermediate to the human hazards. The energy/material transfer or blockage also yield the intermediate correlations. Other causes of the intermediate correlations are (1) false information transfer; (2) correct information blockage; (3) hardware, software, and function failures; and (4) background parameters known as the performance shaping factors, etc.

VI. ENUMERATION OF THE HAZARDS OF THE HUMAN-ROBOT SYSTEM

The human-robot system consists of human, robot, peripheral machines, equipment, tools, works, and other materials. Each system component except for human could be an assailant or an accomplice. The human is always the victim, and in some cases could be the accomplice. Some of the hazard scenarios from the points of view of the character analysis and the causes of ultimate or intermediate correlations are briefly summarized.

A. Energy Transfer

The robot associates with it two types of motions: controlled and uncontrolled. The latter includes an inadvertant release of works from the hands, or a fall on a slope. The human is also mobile and he/she could erroneously crash against the robot. These motions trigger the kinetic energy transfer.

1. ASS-VIC: The robot as an assailant crashes severely against the human, resulting in fatality or injury.
2. ACC-ASS-VIC: The robot as an accomplice incorrectly interferes with a peripheral machine which, in turn, inadvertently falls or moves, resulting in fatality or injury.
3. ACC-VIC-ASS-VIC: The robot as an accomplice pushes the human, forcing him to manipulate incorrectly a peripheral machine which, in turn, hurts the human.
4. ACC-VIC-ASS-VIC: The robot as an accomplice pushes the human away (ACC-VIC), forcing him/her to enter a dangerous zone of a peripheral machine (VIC-ASS) which, in turn, injures the human (ASS-VIC).

Some robots handle hot or cold materials or equipment having heat sinks or sources. Most robots have electric power sources to drive body or arms. The robot and equipment could generate vibrations and noises. Some tools and sensors on the robot generate radiation of various types. Similar hazard scenarios to the kinetic energy cases hold for these energy sources.

B. Material Transfer

Some robots handle poisons, corrosives, chemicals, and radioactive materials and liquids.

1. ASS-VIC: The robot as an assailant spills a chemical over the human, resulting in chemical burns.
2. ACC-ASS-VIC: The robot as an accomplice spills water over an electric circuit of a peripheral machine which, in turn, fails and causes human hazards.
3. ACC-VIC-ASS-VIC: The robot as an accomplice spills a corrosive near the human (ACC-VIC) who, in turn, is forced to enter a dangerous zone of a peripheral machine (VIC-ASS) to avoid the corrosive, resulting in an injury (ASS-VIC).
4. ACC-VIC-ASS-VIC: The same as above except for the human incorrectly manipulates the peripheral machine.

C. Energy or Material Blockage

This is typical when the robot functions as a feeder machine. Only one example is shown: ACC-ASS-VIC. The robot as an accomplice fails to supply lubricants to a peripheral machine (ACC-ASS) which, in turn, fails and hurts the human (ASS-VIC).

D. Information Transfer and Blockage

This is typical when the robot acts as an information carrier. Only two examples are shown.

1. ACC-ASS-VIC: The robot gives incorrect data to a peripheral machine which, in turn, hurts the human.

Table 2
REAL SCENARIOS AND CHARACTER PATTERNS

Phase	Scenario	Character pattern	Frequency
Operation	Something was wrong with the robot. A workman entered the mobile area to inspect it and was hit by the robot arm.	ACC(R)-VIC-ASS(R)-VIC	8
Operation	Something was wrong with a peripheral machine. A workman entered the mobile area of a robot to inspect the abnormality, and was hit by the robot arm.	ACC(P)-VIC-ASS(R)-VIC	2
Operation	A workman erroneously entered a mobile zone of a robot which, in turn, hit him.	VIC-ASS(R)-VIC	2
Startup	An operator in the mobile zone started a robot.	VIC-ASS(R)-VIC	3
Instruction Inspection Repair Alignment	1. An operating robot suddenly moved toward an incorrect direction, hitting a workman.	ASS(R)-VIC	4
	2. An operating robot changed its motion because a workman incorrectly manipulated it. He was hit by the robot.	VIC-ASS(R)-VIC	2
	3. An operating robot hit a workman because he failed to predict the robot motion.	VIC-ASS(R)-VIC	3
	4. An inactive robot suddenly moved and hit a workman when a person inadvertently turned on a start switch.	ACC(H)-ASS(R)-VIC	2

2. ACC-VIC-ASS-VIC: The robot as an accomplice gives incorrect message to the human (ACC-VIC) who, in turn, manipulates a peripheral machine erroneously (VIC-ASS). The machine suddenly moves and hits the human (ASS-VIC).

E. Component Failures
The robot could lose its function when component failures occur.

1. ACC-ASS-VIC: Suppose a robot is driving a carrier. The robot incorrectly controls the carrier which, in turn, crashes against the human.
2. ACC-VIC-ASS-VIC: Suppose that a robot is supporting a person working at a high place. The robot as an accomplice fails and inadvertantly releases the person (ACC-VIC). He falls (VIC-ASS) and crashes against the ground (ASS-VIC).

F. Background Factors
Shapes, structures, weights, sizes, positions, and directions of the robot could increase the occurrence probabilities of hazards. Similarly, difficulties of tasks such as maintenance, installation, or instruction contribute to the hazards. The performance-shaping factors of the THERP technique are useful in identifying what scenarios are more plausible than others, given a set of background factors.

This section concludes by presenting real scenarios which occurred by November 1983.[1] Table 2 lists them together with task phases, frequencies, and character patterns. A total of 26 incidents are involved. In the character pattern, symbols R and P in the parentheses denote robot and peripheral machine, respectively. Table 3 lists typical near misses.

VII. PRELIMINARY HAZARD EVALUATION

The historical data in Tables 1, 2, and 3 indicate that the kinetic energy transfer from

Table 3
EXAMPLES OF TYPICAL NEAR MISSES

No.	Situation and causes	Phase
1	A push button of a teaching box failed, resulting in an uncontrolled motion of a robot manipulator.	Instruction
2	An instructor violated an operation sequence, and a robot manipulator inadvertently returned to its origin.	Instruction
3	A robot manipulator suddenly moved with a high speed when an electric cable was inadvertently cut.	Instruction
4	An operator error forced a robot manipulator suddenly to start moving in a wrong direction.	Instruction
5	A robot manipulator incorrectly moved and crashed against a machine when an instructor turned on the main power switch of the robot.	Instruction
6	A workman approached to the active manipulator when it inadvertently released a work. The manipulator came near to hit him.	Operation
7	A workman got too close to the manipulator when he corrected a position of a work and turned on the main power switch. His head was near to contact the manipulator.	Operation
8	A manipulator approached a workman when he was repositioning a work.	Operation
9	The same as in 8 except for the near miss which occurred during a normal positioning of a work.	Operation
10	A workman was inspecting a manipulator because it moved incorrectly. Another workman erroneously started the manipulator.	Inspection
11	A workman was inspecting a failure of a vacuum holder. The manipulator suddenly moved by the air left in the controller.	Inspection
12	A repairman was aligning a fixed-sequence robot with the main power switch being on. He erroneously touched a limit switch, and the manipulator moved.	Alignment
13	A workman got too close to a manipulator when he intended to remove waste shred without turning off the robot main power switch.	Cleanup

robot to human constitutes major causes of unsafe incidents. This is chiefly due to the fact that most robots have been exploited in manufacturing industries such as automobile makers where critical hazards like explosions, fires, and serious poisonings seldom occur. Hazards other than the kinetic energy transfer type will become important in the future when robots are used in nuclear, chemical, or medical industries, etc. In the following sections, the remaining steps of safety assessment by considering the currently major hazard, i.e., "human is struck by robot", are demonstrated.

VIII. COMPREHENSIVE CAUSAL MODELS

A specific causal model like a fault tree is usually constructed for a hazard as a top event when a particular system to be analyzed is specified in some detail. The construction and validation of such a specific model are two subjects of importance. The comprehensive causal model developed in this section, on the other hand, considers a general situation which contains every aspect of possible human-robot systems. The specific model can be constructed from the comprehensive model by extracting only factors specific to the system under consideration. The comprehensive model is also useful for the validation of the specific model because the former should contain the latter as a special case.

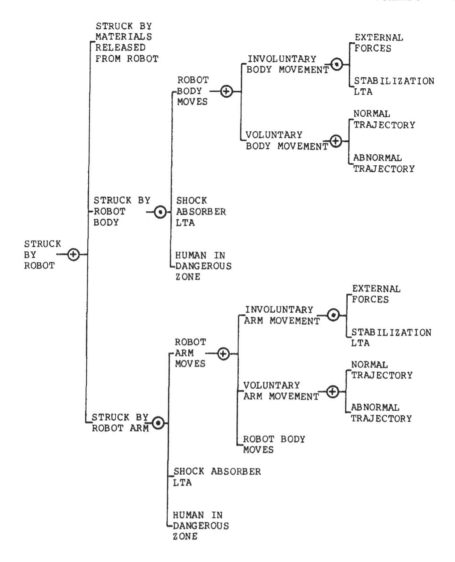

FIGURE 2.   A tree top of a comprehensive causal model.

MORT[6] is a comprehensive hazard model, but too general to be applied to the human-robot system.

Only one type of the top event, i.e., "human is struck by robot" is considered here. Figure 2 shows a tree top of the comprehensive causal model, where symbols + and · in circles denote logic OR and AND, respectively. The top event is developed into an OR combination of the three events:

1.   The human is struck by materials released from the robot.
2.   The human is struck by the robot body.
3.   The human is struck by the robot arm.

The last two events are developed explicitly in Sections VIII.A and B because of the higher plausibilities in practical situations. The first event can be analyzed in a similar way.

A. Human is Struck by Robot Body

As shown in Figure 2, the following three events must exist simultaneously in order for the human to be struck by the robot body.

FIGURE 3.   A development of "external forces".

1. The robot body moves when the human is in a dangerous zone.
2. Shock absorbers are less than adequate (LTA). The absorber mitigates the unwanted energy transfers when the robot is touching the human. In MORT terminology, the shock absorber is a kind of barrier which, in general, can locate on robot, or between robot and human, or on human. Figure 7 shows a typical development of "barrier LTA" event.
3. The human is in the dangerous zone around the robot where the robot movement could cause the fatality or injury.

The robot has its own motive power. We can thus consider two cases of robot movement, as shown in Figure 2: (1) an involuntary movement due to external forces other than the motive power and (2) a voluntary movement caused by the motive power.

1. Involuntary Body Movement

The involuntary movement occurs when a "stabilization LTA" event coexists with an external force of a suitable magnitude. As shown in Figure 3, the external force includes gravity, collision reaction, vibration, earthquake, overload, etc. The robot turns over, falls down, starts moving, or changes direction by involuntary movement.

Figure 4 develops "stabilization LTA" in Figure 2. Two phases are considered: the robot is off-duty and thus not alive, and the robot is on-duty and alive. The off-duty robot is a motionless structure and becomes unstable when "physical stabilizer LTA" in Figure 4 coexists with the external force. The stabilizer includes hardware support, fence, bar, wall, bump, etc.

The on-duty robot, on the other hand, moves continuously or intermittently. As shown in Figure 4, "physical stabilizer LTA" is a cause of "stabilization LTA". For the on-duty robot, there is yet another cause, i.e., "attitude controller LTA" because the robot attitude is usually maintained by feedback control features. This event is, in turn, developed into the following causes, completing Figure 4.

1. Software LTA: This implies defective feedback control software (or algorithm).
2. Software selection error: This implies a use of incorrect software when correct ones are available.
3. Manipulation error: This is a manipulation error when the controller and its software are normal.
4. Actuator failure: The actuator failure is caused by oil contamination, servo valve failure, lack of utilities, etc.,
5. Logic circuit failure: This is caused by disconnection, shorted wire, and other defections.
6. Faulty signal: This includes static electricity, momentary power failure, internal or external noise, internal or external sensing device failure, etc.
7. Others.

In many cases the motion controller in Figure 6 also acts as the attitude controller.

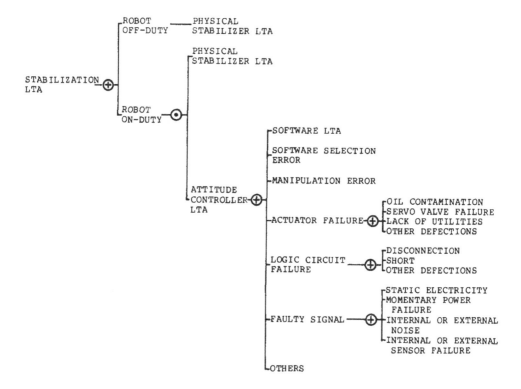

FIGURE 4. A development of "stabilization LTA".

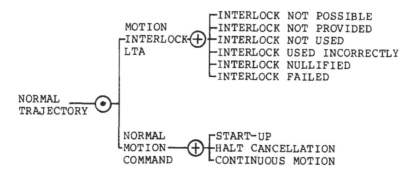

FIGURE 5. A development of "normal trajectory".

2. Voluntary Body Movement

Attention now returns to the voluntary body movement in Figure 2. The robot moves along a normal time-space trajectory within design specifications, or along an abnormal trajectory outside the specifications. These two types of movement are further developed in Figures 5 and 6.

In order for the movement along the normal trajectory to occur, both things must exist simultaneously.

1. Motion interlock LTA: The interlock is a device which detects the human existence along the trajectory, and stops the robot movement accordingly. The interlock must fail in order for the body to strike the human. The failure is developed into interlock not possible, interlock not provided, interlock not used, interlock used incorrectly,

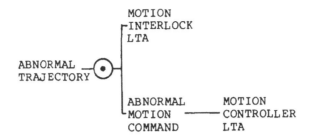

FIGURE 6.   A development of "abnormal trajectory".

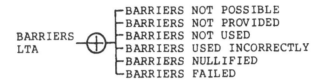

FIGURE 7.   A development of "barriers LTA".

interlock nullified intentionally, and interlock failed. The development is similar to Figure 7 because the interlock and the shock absorber are a kind of "barrier". The interlock is failed by failures of its actuator, sensor, or logic circuit, etc.,

2.   Normal motion command: The normal command is generated by start-up operation or halt cancellation. The halt status is of danger because the robot appears to be off-duty and because the cancellation is usually made automatically.

The movement along the abnormal trajectory is an AND combination of "motion interlock LTA" and "abnormal motion command", as shown in Figure 6. The "motion interlock LTA" can be analyzed in the same way as in Figure 5. The "motion controller LTA" has the same structure as the "attitude controller LTA" in Figure 4.

3. Shock Absorbers LTA

The event, "shock absorbers LTA" in Figure 3 is developed in the same way as Figure 7, resulting in absorbers not possible, absorbers not provided, absorbers not used, absorbers used incorrectly, absorbers nullified intentionally, and absorbers failed.

4. Human In Dangerous Zone

The dangerous zone depends on the mobility of the robot. If the robot can move like an automobile, the zone is large, otherwise, the zone is a small area around the robot. The event "human in dangerous zone" is developed in Figure 8. There are two possibilities of human intrusion into the dangerous zone: authorized intrusion and unauthorized intrusion. Authorized intrusion occurs to perform tasks permitted by job and safety instructions. We can consider four types of the tasks: to serve robot, to serve peripheral machine, to be served by robot as in medical applications, and to be served by peripheral machine. The first and second types include transportation, installation, alignment, instruction or teaching, test, start-up, steady operation, troubleshooting, and maintenance and repair.

Unauthorized intrusion is an AND combination of "intrusion barriers LTA" and an initiation of the intrusion. Barriers such as warnings, bars, doors, etc. are usually provided to prevent the unauthorized intrusion. They must fail. The event, "intrusion barriers LTA" has the same structure as in Figure 7. The initiation of the unauthorized intrusion is developed

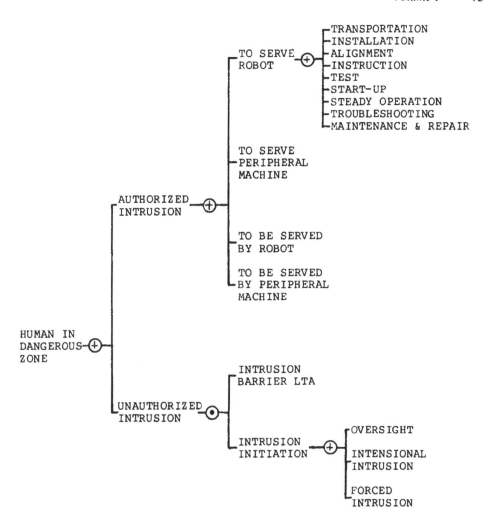

FIGURE 8. A development of "human in dangerous zone".

into oversight, intentional intrusion, and forced intrusion. The forced intrusion typically occurs when a peripheral machine pushes the human toward the robot, or the human stumbles.

Note that we have identified five types of barriers.

1. Shock absorber: This prevents the robot body from hurting the human when a collision occurs.
2. Stabilizer: This stabilizes or stops the robot body movement caused by external forces, and includes physical stabilizers and attitude controllers.
3. Motion interlock: This stops the voluntary body movement before the collisions occur.
4. Motion controller: This maintains the robot body movement along the normal trajectory, and includes feedback controllers and mechanical stoppers.
5. Intrusion barrier: This prevents the unauthorized intrusion into the dangerous zone.

B. Human is Struck By Robot Arm

This event can be developed similarly to the body case. Only one exception is that the arm movement can also be caused by the body movement, as shown in Figure 2. The robot body movement here is developed in the same way as in Section VIII.A.

IX. PHASE DECOMPOSITION AND MIN CUT SETS

The comprehensive causal model developed in Section VIII describes the top event occurrence during a variety of phases ranging over different time intervals. Consequently, some events in the model are not statistically independent of each other. For example, "off-duty robot" and "on-duty robot" in Figure 4 are mutually exclusive. Similar exclusiveness also holds for "startup" and "halt-cancellation" in Figure 5, and for "transportation", "installation", "alignment", "instruction", etc. in Figure 8.

Occurrence probabilities of events vary according to the phases of the human-robot system. For instance, the voluntary body movement frequently occurs during test, startup, and steady operations, but seldom occurs during transportation or installation when the robot is off-duty.

The statistical dependence and the occurrence probability variation are a source of difficulty to quantify the comprehensive causal model. We must thus apply a divide-and-conquer strategy to the model, i.e., the phase decomposition.

Three types of factors are available for determining a particular phase:

1.   Robot life cycle characteristics: Manufacture, transportation, installation, alignment, arrangement, instruction, test operation, startup, steady operation, troubleshooting, maintenance and repair, abolition.
2.   Robot motion characteristics: Involuntary movement of off-duty robot, involuntary movement of on-duty robot, voluntary, normal movement, and voluntary, abnormal movement.
3.   Human intrusion characteristics: Authorized intrusion and unauthorized intrusion.

Each phase is now described by a three-dimensional vector consisting of the three characteristics. Consequently, we have a total of $10 \times 4 \times 2 = 80$ phases for the body or the arm, neglecting manufacture and abolition. Some of them are more plausible than others. For instance, the phase transportation, involuntary movement of off-duty robot, authorized intrusion is more conceivable than the phase transportation, voluntary, normal movement, authorized intrusion. The comprehensive causal model is greatly simplified and becomes tractable, given a particular phase.

These are the four phases frequently observed at the steady operation of robot.

Phase A
1.   The robot is under steady operation.
2.   The robot moves voluntarily and normally.
3.   Authorized intrusion of human.

Phase B
1.   The robot is under steady operation.
2.   The robot moves voluntarily and normally.
3.   Unauthorized intrusion of human.

Phase C
1.   The robot is under steady operation.
2.   The robot moves voluntarily and abnormally.
3.   Authorized intrusion of human.

Phase D
1.   The robot is under steady operation.
2.   The robot moves voluntarily and abnormally.
3.   Unauthorized intrusion of human.

Given one of these phases, a simplified version of the causal model is obtained by trimming events irrelevant to the phase. Minimal (min) cut sets are identified accordingly. Let us consider the event, "human is struck by robot body". Typical min cut sets are listed below.

Phase A

- *Cut 1 (halt cancellation, motion interlock LTA, shock absorber LTA, workman for steady operation):* For simplicity of description, some of the events in the minimal cut set are not developed completely. For instance, "shock absorber LTA" could be replaced by one of the more basic events. Event "motion interlock LTA" has the same structure as Figure 7, and similar replacements hold. The halt cancellation occurs for the robot which moves and stops intermittently at the steady operation.

- *Cut 2 (continuous motion, motion interlock LTA, shock absorber LTA, workman for steady operation):* Assume that the workman is well informed of the time-space trajectory of the robot motion. Assume also that the robot automatically moves and stops. In this situation, cut 1 is more important than cut 2 because it is conceivable that the workman fails to notice the automatic halt cancellation. Cut 1 seems to occur frequently during an early time period of the intrusion since the workman may not yet be familiar with the intermittent behavior of the robot. Audio and visual warnings must be provided to notify the halt cancellation. In addition, shock absorbers such as helmet and protector are required. As shown in Figure 7, the workman may not use the shock absorbers when available. Similarly, he may nullify the motion interlocks. This occurs frequently when the workman is encumbered with these protective features. If steady operation requires no workman, these two cut sets do not occur.

- *Cut 3 (startup, motion interlock LTA, shock absorber LTA, workman for steady operation):* This cut does not hold because the steady operation excludes the startup.

Phase B

- *Cut 4 (halt cancellation, motion interlock LTA, shock absorber LTA, intrusion barrier LTA, unauthorized intrusion):* The unauthorized intrusion was developed in Figure 8. The cut set is similar to cut 1 except for the existence of "intrusion barrier LTA" which indicates access control failures.

- *Cut 5 (continuous motion, motion interlock LTA, shock absorber LTA, intrusion barrier LTA, unauthorized intrusion):* This cut is typically observed when a peripheral machine forces a person to come closer to an operating robot.

Phase C

- *Cut 6 (halt cancellation, motion interlock LTA, motion controller LTA, shock absorber LTA, workman for steady operation):* The "motion controller LTA," which can be developed similarly to the "attitude controller LTA" in Figure 4, causes the abnormal motion of phase C. This is typically observed when a momentary power failure occurs during the halt mode; the servo valve then has a wrong position when halt cancellation command is received after the power recovery. The robot moves abnormally and may strike the workman.

- *Cut 7 (continuous motion, motion interlock LTA, motion controller LTA, shock absorber LTA, workman for steady operation):* The robot moves abnormally along a wrong space trajectory or in an incorrect timing, and may strike the workman.

Phase D

- *Cut 8 (halt cancellation, motion interlock LTA, motion controller LTA, shock absorber LTA, intrusion barrier LTA, unauthorized intrusion)*

- *Cut 9 (continuous motion, motion interlock LTA, motion controller LTA, shock absorber LTA, intrusion barrier LTA, unauthorized intrusion):* Cuts 8 and 9 are similar to cuts 6 and 7 except for the involvement of "intrusion barrier LTA".

X. A QUANTITATIVE ANALYSIS

A. System Description

Consider a work station consisting of a fixed-body robot and an N.C. machine. The station is surrounded by a wire fence which has a door for operator entrance and exit. The robot picks up a work on a belt conveyor, supplies it to the N.C. machine, and returns the product to the conveyor. The operator must unlock the door to enter the workstation. The robot can be inactivated in two ways: the operator turns on a stop switch outside the fence and a sensor at the door senses the operator entrance, inactivating the robot automatically. Therefore, there are two motion interlocks, one is manual and the other is automatic.

The operator enters the work station for maintenance, inspection, alignment, instruction (teaching), etc. He also enters there when the robot erroneously positions the work or product. The door key is exclusively possessed by the operator, and chances of unauthorized intrusions by other persons are relatively rare. Also neglected is the event, "struck by robot body" because the robot body is fixed. In the following section, the event, "operator is struck by robot arm" is quantitatively analyzed.

B. Operator is Struck by Robot Arm

A typical cut set is similar to cut 1 in Section IX: halt cancellation, motion interlock LTA, shock absorber LTA, operator is inside the fence. It is assumed that none of the shock absorbers is available. As described in Section A, two types of motion interlock are provided: manual and automatic. For simplicity it is supposed that the manual motion interlock becomes LTA if and only if the operator forgets to turn on the stop switch outside the fence. It is also assumed that the automatic motion interlock fails only if the door sensor fails. Thus, the cut set can be rewritten as (E1, E2, E3, E4) where:

E1:   Halt cancellation. This occurs when the N.C. machine finishes manufacturing the last work and the robot arm starts moving to handle the product and the next work. The robot arm stands still during the other time intervals.

E2:   Operator fails to turn on the stop switch when he enters the work station.

E3:   Sensor failure of the automatic motion interlock.

E4:   Operator is in a hazardous zone inside the fence.

C. Basic Event Data

Event E1. Assume that the work station can produce a_1 products per hour. Then, the halt cancellation command to the robot occurs a_1 times per hour. Assume also that the robot requires $1/b_1$ hr to handle a pair of product and work. The event E1 can be modeled by an exponential distribution process with failure rate a_1 and repair rate b_1. The existence probability of event E1 coincides with the unavailability of the exponential process.

Event E2. This is a human error and can be modeled by an inhibit gate with a constant probability q_2 of occurrence.

Event E3. Assume a nonrepairable door sensor. The event can be quantified by an exponential process with failure rate $a_3 > 0$ and repair rate $b_3 = 0$.

Event E4. Assume that the robot arm erroneously positions the work or product a_4 times per hour. Assume also that the operator stays in the dangerous zone inside the fence $1/b_4$ hour to deal with the work or the product. Then the event is modeled by failure rate a_4 and repair rate b_4.

D. Quantification by KITT

The KITT (Kinetic Tree Theory) yields, among other things, the following probabilistic parameters for basic events, cut sets, and the top event of a fault tree:

<table>
<tr><th colspan="2">Table 4
BASIC EVENT DATA</th></tr>
</table>

Event E1	$a_1 = 30$,	$1/b_1 = 1/360$
Event E2	$q_2 = 1/100$	
Event E3	$a_3 = 1/8760$,	$b_3 = 0$
Event E4	$a_4 = 1/24$,	$1/b_4 = 1/36$

Table 5
EXPECTED NUMBER OF
INCIDENTS PER YEAR

Manual & auto	Manual	Auto
1.11	3.08	111

Table 6
INSPECTION INTERVALS AND
EXPECTED NUMBERS OF INCIDENTS
PER YEAR

Inspection interval	Manual & auto	Manual	Auto
1 hr	0.000175	3.08	0.0175
1 day	0.00419	3.08	0.419
1 month	0.120	3.03	12
2 months	0.235	3.03	23.5
6 months	0.642	3.03	64.2
1 year	1.11	3.08	111

1. Unavailability: This is an existence probability.
2. Failure intensity: An expected number of occurrences per unit time at time t.
3. Integral characteristic: Expected number of occurrences until time t.

We can derive the other two cases from the cut (E1, E2, E3, E4).

1. The manual motion interlock is not provided. The cut set becomes (E1, E3, E4).
2. The automatic motion interlock is not provided. The cut set becomes (E1, E2, E4).

Suppose the basic event data in Table 4. The halt cancellation command occurs every 2 min, and the robot is active for 10 sec. The operator forgets to turn on the stop switch once per 100 instrusions. The mean time to failure of the door sensor is 1 year. The robot arm fails to position the product or work every day. The operator requires 100 sec to correct it. A computer program called SUPERPOCUS yields the result in Table 5. We see that the best cut set (E1, E2, E3, E4) occurs 1.11 times per year. The combination of manual and automatic interlocks reduce the incidents, but it is still unsafe because the door sensor is not inspected. Table 6 shows the effects of the inspection intervals of the door sensor. We see that a 1 month inspection interval reduces the incidents about ten times compared with the nonrepairable case. If the number 0.120/year is still unsatisfactory, a multisensor door system with one month inspection interval will considerably improve the system safety.

In quantifying a cut set, the KITT considers all sequences of basic events in the cut set. There are cases, however, that only some sequences are feasible, and hence the KITT yields conservative values. For instance, consider the cut set without the manual motion interlock (E1, E3, E4). A natural sequence for the cut set is (1) the door sensor fails (E3), (2) the operator enters the work station and stays there (E4), and (3) halt cancellation occurs (E1). Sequences such as E4, E3, E1 have no practical meaning because the sensor remains normal at the operator entrance, and hence the robot can be inactivated. If more accurate quantification than KITT is required, then the sequence can be quantified by a convolution integral or a Markov analysis.

REFERENCES

1. **Mochizuki, S.,** Introduction of Industrial Robots and Safety Countermeasures, Japan Labor Total Research Institute Co., Ltd., Tokyo, 1984 (in Japanese).
2. **Vesely, W. E.,** A time-dependent methodology for fault tree calculation, *Nucl. Eng. Design,* 13, 337, 1970.
3. **Henley, E. J. and Kumamoto, H.,** *Reliability Engineering and Risk Assessment,* Prentice-Hall, Englewood Cliffs, N.J., 1981.
4. **Sato, Y. and Inoue, K.,** Safety assessment of human-robot systems. I. Hazard identification based on the action-changes and action-chains model, *Trans. Jpn. Soc. Mech. Eng.,* 51(468-C), 1985.
5. **Sato, Y., Inoue, K., and Kumamoto, H.,** Safety assessment of human-robot systems. II. Logic models for the analysis of the accident causation mechanisms, *Trans. Jpn. Soc. Mech. Eng., Ser. C,* 52(474), 823, 1986.
6. **Johnson, W. G.,** *MORT Safety Assurance Systems,* Marcel Dekker, New York, 1980.
7. **Swain, A. D. and Guttman, H. E.,** Handbook of Human Reliability Analysis With Emphasis On Nuclear Power Plant Applications, NUREG/CR-1278, U.S. Nuclear Regulatory Commission, Washington, D.C., 1983.

Chapter 5

PLANT FAULT DIAGNOSIS EXPERT SYSTEM BASED ON PC DATA MANIPULATION LANGUAGES

Hiromitsu Kumamoto

TABLE OF CONTENTS

I. INTRODUCTION

In this day of computerization many problems still exist that are intractable by ordinary computer programs, and must be left up to domain experts. The expert system (ES) originated from a conjecture that computers could exhibit much higher levels of problem-solving capabilities if they exploited the domain expert knowledge. This perspective is now changing into a sure confidence via a variety of past applications to fields including medicine, engineering, mathematics, law, education, language, etc. These applications have yielded many expert systems of various types, and some of them are even commercially accessible.[1-3] Famous systems include MYCIN for infection disease diagnosis, DENDRAL for chemical structure elucidation, and PROSPECTOR for mineral deposit exploration.[2]

Most ESs have been programmed in LISP-like languages.[4] The LISP is suitable for the expert knowledge exploitation because it can manipulate lists, and invoke procedures recursively. Some people believe that PROLOG is a promising language for expert system programming and knowledge representation.[5,6] The Japanese fifth generation computer project is using PROLOG as a core language.

Expert knowledge must be represented in computer-readable forms before its exploitation. The most typical representation scheme is based on a collection of IF-THEN rules. As a matter of fact, the notable MYCIN utilizes this type of knowledge representation. The collection, which is called a knowledge base by the expert system convention, can be regarded as a database: in a most simplistic view, such a database consists of IF- and THEN-part fields or attributes. Knowledge exploitation, on the other hand, can be considered as a sequence of information retrievals and database modifications. Consequently, it becomes an interesting attempt to employ existing database technologies for the knowledge representation and exploitation.

A so-called inference is a type of knowledge exploitation. Basically, the inference algorithm monitors the IF-parts, recursively fires rules applicable, and makes conclusions. The expert system, however, requires a variety of knowledge in addition to the IF-THEN rules used by the inference algorithm. For instance, it requires knowledge to justify the rules. This increases the credibility of the inference made by the expert system. The ES application to the plant fault diagnosis, among other things, must make use of information such as: (1) maintenance and repair records, (2) records of abnormal occurrences, (3) methods of confirming facts about plants, (4) design specifications and location of hardware, (5) hazardous material information, (6) feasible ranges of process variables, (7) plant recovery measures following fault diagnosis, (8) other manuals and documents, etc. The diversity of the information suggests that the resultant knowledge base would be created, modified, and exploited most suitably by the database technology.

This chapter proposes programming the expert system for plant fault diagnosis in a data manipulation language (DML) of a relational database.[7,8] The dBASE or K-MAN language on a personal computer (PC) is used as the DML.[9-11] More reasons for using the DML are presented in Section II.

A number of studies have been conducted to apply the expert systems to plant fault diagnoses.[14-17] Most of them are programmed in artificial intelligence languages such as LISP or PROLOG, and a few are programmed in the DML. An alternative approach to the fault diagnosis is based on cause consequence diagrams, and this is not specifically related to the expert system.[12]

This chapter is organized as follows. Section II presents more reasons for using the DML rather than LISP or PROLOG. Section III introduces a knowledge acquisition and representation scheme of incremental refinement type. In this scheme, the plant states are successively decomposed into more definitive states as new observations are brought to light. Each observation denotes a clue the expert looks at during the fault diagnosis. The incremental refinement facilitates the knowledge acquisition from the plant. It also allows the utilization of checklists and fault trees as knowledge sources other than the human experts themselves. The resultant expert knowledge is visible as an AND/OR tree.

Section IV demonstrates two sets of IF-THEN rules obtained by a first engineer for two ship engine systems. The reader can see how the expert locates causes through the incremental refinement scheme.

Section V clarifies structures of a knowledge base consisting of three databases. IF-THEN rules and other information such as rule justifications, fact confirmation methods, and plant recovery measures are stored in the databases supervised by the DML.

In Section VI, inference, explanation, and other modules of the expert system are programmed as an application software which creates, modifies, and accesses the databases. Main features of the PC-based expert system are reviewed. Section VII demonstrates the ES through applications to the fault diagnosis of the ship engine cooling system.

II. EXPERT SYSTEM AND RELATIONAL DML

As described in Section I, most expert systems have been programmed in artificial intelligence languages such as LISP or PROLOG. The use of the relational DML must be justified here.

A. Information Retrieval Capability

It is well known that the relational database is based on Codd's theory which states, in rough terms, that any information explicitly or implicitly stored in the database can be thoroughly retrieved and exploited by the DML, satisfying a certain condition called a relational completeness.[7,8] As a matter of fact, the Japanese fifth generation computer project is implementing PROLOG by a hardware DML of a relational database machine with parallelisms. This is a reasonable implementation because PROLOG is primarily a data retrieval language where the retrieval conditions are described in terms of predicate logic. DML is a more basic and potentially more powerful language than LISP or PROLOG, provided that the problem under consideration is a database application.

The expert system for the plant fault diagnosis can be formulated as a database application program, and a natural consequence is to program the ES by the DML. The relational completeness of the PC-based DML is still doubtful. However, the DML of K-MAN is a well-organized subset of a relationally complete language called SQL. The DMLs for the personal computers are a challenging field and the languages are being renewed almost every day, as indicated by the recent dBASE III release in the U.S.

B. Database Retrievals for Expert System

The panel discussion at the 1983 IJCAI (International Joint Conference on Artificial Intelligence) pointed out the merits of database retrievals for industrial ESs.[13] A DNA database is cited as an example to be retrieved. As described in Section I, the plant fault diagnosis involves a variety of information and the database retrieval is vital to the ES. For instance, explanation capability of the ES is improved considerably by timely accesses of key information.

The ES usually solves problems through dialogue between the computer and the computer operator. The computer asks the operator for data missing from its internal databases and the operator retrieves relevant information from the brain or external databases, providing the computer with the data. This type of dialogue naturally leads to the ES programmed by the DML. As shown in Section V.B, the inference capability of the ES can be viewed as data modification and retrieval coupled with an AND/OR tree.

C. Artificial Intelligence and Database Technology

The ES belongs to a branch called artificial intelligence that has developed almost independently of the database discipline. The two branches are still isolated from each other, and the same or similar concepts are called by different names. For instance, the "field" of the database terminology corresponds to the "slot" of the artificial intelligence; the semantic network and the frame in the artificial intelligence are special types of database; index files can improve the processing speed of ES, but such a technique has been used frequently in the database discipline.

System configurations are also similar to each other. The expert system consists of inference and explanation modules, knowledge bases, and a working memory called a blackboard. The database system is composed of application programs, databases, and working memory. The notable, mutual independence between the inference module and the knowledge base corresponds to the one between an application program and the databases.

In this chapter the knowledge bases are implemented as the databases, and the inference, explanation, and other modules are written as application programs to the databases. This offers a chance of mediating between the two isolated disciplines, i.e., the artificial intelligence and the database technology.

D. Language Practice

A criterion for adopting a language is how people become used to it. The higher level languages do not necessarily prevail more. FORTRAN is still widely used despite the releases of new, higher level languages such as PASCAL. DML is the most familiar language for people in the database discipline to write application programs. The people need not learn LISP nor PROLOG if the ES can be programmed by DML. It is worth clarifying the current DML capability to implement the ES.

E. Language Understandability

There is often difficulty in understanding LISP or PROLOG expressions because of the notable use of parentheses. For example, consider the PROLOG expression:

$$[(\text{Bill } 53)(\text{Jane } 47)) \text{ parent-of } ((\text{Jim } 17)(\text{Karen } 15)] \qquad (1)$$

The implication is not so obvious. Numbers are possibly ages, but they may be weights or birth years. Possibly, the parents are Bill and Jane, and the children are Jim and Karen, but the converse is also plausible. Only English name conventions suggest that Bill or Jim is the father, and that Jane or Karen is the mother.

An equivalent database expression to Equation 1 is

father name: Bill,	father age: 53
mother name: Jane,	mother age: 47
first child name: Jim,	first child age: 17
second child name: Karen,	second child age: 15 (2)

This is obviously more understandable than the PROLOG expression (Equation 1).

Descriptions about the data themselves have been important topics in the database discipline.[7,8] The PC-based DMLs inherit some of the core data description schemes from the mainframe. The relational DMLs adopt tabular forms like spread sheets, thus increasing the understandability. The DML of K-MAN or dBASE is a structured language and as comprehensible as BASIC.

F. Database Creation and Modification

The DML is the best tool for creating and modifying information systems including knowledge bases. The modification features are used effectively for incremental knowledge revision such as TEIRESIAS.[2]

G. Procedural and Declarative Languages

There is still a big debate about the merits and demerits of procedural and declarative languages.[2] The DML of dBASE or K-MAN is procedural, and has program flow control statements such as WHILE-DO, IF-THEN-ELSE, and CASE which are indispensible to structured languages. PROLOG, on the other hand, is a typical declarative language where expressions are compact but program flows are hard to control. For instance, a goal-oriented search can be implemented quite naturally by PROLOG, but a data-oriented search seems to require some tricks. In a simple goal-oriented search implementation, program statements must be explicitly rearranged to change a validation sequence of goals. PROLOG may be suitable for sophisticated artificial intelligence applications such as theorem prover where complicated program controls must be left up to the computer. For the MYCIN type fault diagnosis, however, procedural languages are more suitable than declarative languages because people can control the program flows for better implementation of ES. The LISP has procedural statements, and has been used successfully in the artificial intelligence discipline, including ES and symbolic manipulations. This language, however, requires a great deal of expertize to become used to it.

H. Processing Speed

Frequent disk accesses, in particular, considerably slow the processing speed of DMLs of current PCs. Hardware and software improvement is surely resolving this difficulty.

Currently, a so-called RAM disk can be used effectively for speeding up personal computer DML applications.

III. FAULT DIAGNOSIS KNOWLEDGE

This section is concerned with a knowledge acquisition and representation scheme for the plant fault diagnosis. An incremental refinement scheme yields a set of IF-THEN rules which constitutes a core of the fault diagnosis knowledge. Other knowledge such as fact confirmation methods and plant recovery measures, etc. can be identified easily once the core has been established. The incremental refinement scheme is particularly suitable for ES applications to the plant fault diagnosis.

First, consider the following PROLOG rules[5,6]:

$$x \text{ should-take } y \text{ if}$$

$$x \text{ complains-of } u \text{ and}$$

$$y \text{ suppresses } u \text{ and}$$

$$\text{not } y \text{ is-unsuitable-for } x \tag{3}$$

$$y \text{ is-unsuitable-for } x \text{ if}$$

$$y \text{ aggravates } v \text{ and}$$

$$x \text{ suffers-from } v \tag{4}$$

These rules express a general constraint among patient x, medicine y, primary disease u, and secondary disease v. The constraint resembles with the Newton equation of motion by which physical variables such as mass, force, and acceleration are constrained. It is unrealistic to expect that the fault diagnosis knowledge could be acquired and represented in forms of general constraints such as PROLOG rules.

This chapter defines the fault diagnosis as a process which decomposes the plant state into more definitive states as new observations are brought to light and identifies causes of fault by most definitive states at the ultimate decomposition. The plant state in general implies a set of causes. The fault diagnosis locates a cause when it reaches a plant state containing a single element. This type of fault diagnosis is an incremental refinement, and is similar to the medical diagnosis of MYCIN.

Each observation to decompose (or refine) the plant state denotes evidence that the expert examines. For instance, a first engineer exploits sequences of artful observations for a fault diagnosis triggered by the event "failure to start a ship engine". Such a sequence contains possibilities of eingine air-run, main lever positions, cam shaft positions, etc. The expert observations and resultant rules are described in detail in Section IV.

Each stage of the incremental refinement can be expressed by the following rule:

IF   [plant state i]

AND   (observable fact i_1)

AND ... AND (observable fact i_n)

THEN [plant state j]   (5)

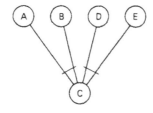

FIGURE 1. A simple AND/OR tree.

In other words, plant state i is refined into more definitive state j, given the observable facts at state i. The fact part, i.e., the point of expert observation, can be an AND combination of two or more facts. The convention used here is to describe the state in square brackets and the observable fact in parentheses. We sometimes use angle brackets to denote explicitly the state at the ultimate refinement.

Consider, as an example, a double-loop cooling system: coolant loop A is cooled by loop B via a heat exchanger. Typical rules are

IF   [Enough coolant circulates in loop A]

AND   (Temperature gauge reading of loop A at

          exit of heat exchanger is high)

THEN [Heat exchanger is ineffective]   (6)

IF   [Heat exchanger is ineffective]

AND   (Temperature guage reading of loop B at

          exit of heat exchanger is low)

THEN <Fouled heat exchanger and hence poor heat

          tranfer is responsible for plant overheat>   (7)

The concluding part of Equation 7 is an ultimate state refinement, and thus a cause of the event, "plant overheat". Equation 6, on the other hand, concludes an intermediate state which must be developed into more basic causes such as the heat exchanger malfunction as in Equation 7, and valve or pump failures of loop B.

Equation 5 is an operator which creates a more specific state from a more general state. The incremental refinement process is equivalent to a collection of these rules, and the resultant expert knowledge is visualized by a hierarchical graph called an AND/OR tree. Figure 1 is a simple AND/OR tree representing two rules "**IF A AND B THEN C**" and "**IF D AND E THEN C**". The arc which crosses two branches A-C and B-C denotes logic AND, i.e., A AND B. It can be seen in Figure 1 that two ANDs flow into node C. These inflows are related by logic OR. Figure 2 is an AND/OR tree which represents the expert knowledge for the engine cooling system described in Section IV.A.

The root of the AND/OR tree is the event to be diagnosed. The knowledge acquisition based on the incremental refinement is a top-down process which is similar to a fault tree (FT) synthesis. Thus, most of the heuristic guidelines for the FT synthesis can apply to the

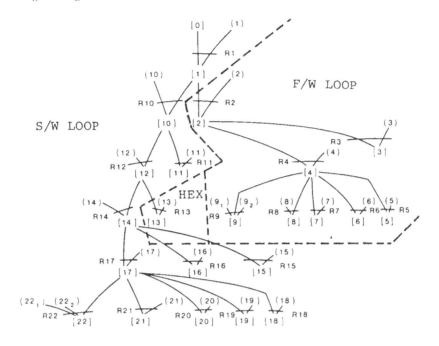

FIGURE 2.   AND/OR tree of 22 rules for ship engine cooling system.

fault diagnosis knowledge acquisition. For instance, a component failure can be developed into defective component such as stuck valve, faulty command to the component, and component utility or resource faults such as power failure or abnormal clock. Another component may be responsible for the faulty command or the utility/resource failure, and the component failure decomposition is iterated until we reach known boundary conditions.

A good checklist has a hierarchical structure similar to the AND/OR tree. Checklists are rarely described in terms of general constraints like Equations 3 and 4. The expert knowledge contained in the well-organized checklists can be directly converted into the rule bases of the incremental refinement type.

The knowledge visualization by the AND/OR tree is of importance. Some expert systems consider randomly arranged rules that are linked dynamically only at the execution. This is known as the modularity of rule-based systems where each IF-THEN rule can be modified independently of others. Our experience shows that such a modularity does not apply to the fault diagnosis rules. It is even harmful to persist on the mutal independence of the rules.

The hard-wired AND/OR tree representation (see Figure 2) is useful for recognizing the whole image of the expert knowledge. Each rule is now conditioned by downstream paths from the root node if a state is to be developed further. For instance, look at Equations 6 and 7. The conclusion of Equation 6 is conditioned by its prerequisites. This allows for the conclusion of the fouled exchanger by Equation 7. It would be difficult to derive the conclusion if only Equation 7 was examined. We can identify Subsystems such as coolant loops and heat exchangers in the hard-wired AND/OR tree of Figure 2 can be identified, and the explanation programs can exploit the results.

In the author's view, the expert knowledge core, i.e., the rule base, is a structured collection of IF-THEN rules, rather than a random arrangement of independent rules. This viewpoint suggests that a more elaborate knowledge representation must exploit a structured language scheme consisting of nested IF-THEN-ELSE-ENDIF blocks, which will be briefly discussed in Section IV.A.3.

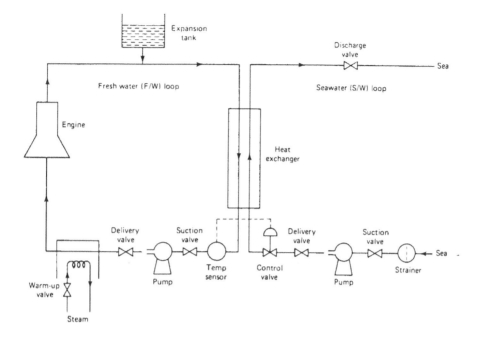

FIGURE 3. Cooling system of ship engine.

IV. TWO EXAMPLES OF RULES BASES

Many expert systems have been proposed thus far, but few details of expert knowledge bases are accessible in journals and books. Presented here in detail are two examples of rule bases constructed by a first engineer for ship engine systems. Rule bases of practical systems and system schematics are useful because (1) people can recognize what the expert knowledge looks like, and (2) different expert systems can be tested by using a common rule base, (3) various suggestions are obtained for improving expert system architectures, and (4) knowledge acquisition processes are clarified and formalized. Part A describes a rule base for a ship engine cooling system, while Part B describes a rule base for a startup system.

A. A Ship Engine Cooling System
1. System Overview
Consider a double-loop cooling system of the ship engine shown in Figure 3. At the engine startup, the warmup stream valve opens and the fresh water is heated by the steam. At steady-state operation of the engine, the valve which controls the steam flow is closed and the engine is cooled by the fresh water loop, which in turn is cooled by sea water through the heat exchanger. The expansion tank is used to replenish the fresh water coolant. Sea water is pumped from the sea into the loop through the strainer which removes foreign materials such as seaweed and fish. The control valve is provided to control the flow of sea water in accordance with the temperature of the fresh water loop. Sea water is discharged to the sea through the discharge valve.

The cooling system consists of four subsystems: (1) the engine; (2) the fresh water loop composed of warmup valve, coolant pump, delivery valve, suction valve, and expansion tank; (3) the heat exchanger; and (4) the sea water loop comprising of strainer, coolant pump, delivery valve, suction valve, flow control valve, temperature sensor, and discharge valve. We may consider that the temperature sensor belongs to the sea water loop because the sensor together with the control valve regulates the loop flow. These subsystems were identified in the AND/OR tree of Figure 2.

As an example, consider what happens when the strainer is plugged at steady-state engine operation. The flow of sea water decreases and loss of coolant occurs in the sea water loop. This causes insufficient heat removal from the fresh water loop in the heat exchanger. The temperature of the fresh water loop increases and the engine starts overheating. The blocked strainer thus becomes a cause of engine overheat.

The abnormal event which triggers the investigation of cause is the engine overheat. For simplicity we assume that the engine itself does not become a cause. Part A.2 describes an incremental refinement process which reveals 16 possible causes including the blockage of the strainer:

1.   A fresh water coolant pump stoppage
2.   A suction valve closure of the fresh water coolant pump
3.   A delivery valve closure of the fresh water coolant pump
4.   A power decrease in motor of the fresh water coolant pump
5.   Impeller damage in the fresh water coolant pump
6.   The warmup valve being left open
7.   Loss of fresh water in the expansion tank
8.   A fouled heat exchanger, i.e., low heat exchange rate
9.   A sea water coolant pump stoppage
10.   A sea water control valve closure
11.   A sea water delivery valve closure
12.   A suction valve closure of the sea water coolant pump
13.   A strainer blockage
14.   A delivery valve closure of the sea water coolant pump
15.   A power decrease in motor of the sea water coolant pump
16.   Impeller damage in the sea water coolant pump

The problem is to extract the one or more causes which fit the plant data, i.e., observable facts.

2. IF-THEN Rules for a Ship Engine Cooling System
The following is a description of the rules used for isolation of the cause, given the event engine overheat. The rule base was constructed by a first engineer.

Rule 1:   [0] AND (1) = = = > [1].
    IF   [Engine overheats]
    AND (Engine itself is not the cause)
    THEN [Cooling system is responsible for overheat]

The conclusion is more specific than "engine overheats" because it is now clear that the cause is in the cooling system. If the fact "engine itself is not the cause" were not directly observable, yet another set of rules to prove or disprove the fact would be required. The expert then looks at the fresh water pressure to examine the amount of coolant in the loop (see Rules 2 and 10; see also Figure 2).

Rule 2:   [1] AND (2) = = = > [2].
    IF   [Cooling system is responsible for overheat]
    AND (Pressure gauge reading of fresh water is low)
    THEN [Loss of coolant occurs in fresh water loop]

The ultimate cause of the loss of coolant is still unknown. The expert next looks at the fresh water coolant pump status.

Rule 3: [2] AND (3) = = = > ⟨3⟩.
    If [Loss of coolant occurs in fresh water loop]
    AND (Fresh water coolant pump is not operating)
    THEN ⟨Fresh water coolant pump stoppage is responsible for overheat⟩

The conclusion in the angle brackets is an ultimate refinement of the plant state and is a possible cause of the engine overheat.

Rule 4: [2] AND (4) = = = > [4].
IF [Loss of coolant occurs in fresh water loop]
AND (Fresh water coolant pump is operating)
THEN [Loss of coolant occurs in fresh water loop with pump operating]

This conclusion is more definitive than "loss of coolant occurs in fresh water loop", the state in the prerequisite. Rules 5 through 9 reveal five causes of the loss of coolant in the fresh water loop with the coolant pump operating. The expert searches for the causes, looking along the fresh water loop.

Rule 5: [4] AND (5) = = = > ⟨5⟩.
    IF [Loss of coolant occurs in fresh water loop with pump operating]
    AND (Pressure gauge reading at suction valve of fresh water coolant pump is low)
    THEN ⟨Closed suction valve of fresh water coolant pump is responsible for overheat⟩

This conclusion is an ultimate cause of the engine overheat.

Rule 6: [4] AND (6) = = = > ⟨6⟩.
    IF [Loss of coolant occurs in fresh water loop with pump operating]
    AND (Pressure gauge reading at delivery valve of fresh water coolant pump is high)
    THEN ⟨Closed delivery valve of the fresh water coolant pump is responsible for overheat⟩

The suction or delivery valves are often left closed after inspection or repair of the coolant pump.

Rule 7: [4] AND (7) = = = > ⟨7⟩.
    IF [Loss of coolant occurs in fresh water loop with pump operating]
    AND (Ammeter reading of motor of fresh water coolant pump is abnormal)
    THEN ⟨Power decrease in motor of fresh water coolant pump is responsible for overheat⟩

The motor may have a wiring failure.

Rule 8: [4] AND (8) = = = > ⟨8⟩.
    IF [Loss of coolant occurs in fresh water loop with pump operating]
    AND (Level gauge reading of expansion tank is low)
    THEN ⟨There is a low water level in expansion tank and bubbles in fresh water loop are responsible for overheat⟩

Rule 9:   [4] AND (9_1) AND (9_2) = = = > $\langle 9 \rangle$.

      IF   [Loss of coolant occurs in fresh water loop with pump operating]

      AND   (Pressure gauge reading at delivery valve of fresh water coolant pump is low)

      AND   (Pressure gauge reading at suction valve of fresh water coolant pump is normal)

      THEN ⟨Impeller damage to fresh water coolant pump is responsible for overheat⟩

The prerequisite of this rule has the **AND** combination of two observable facts which implies the impeller damage. We have finished the incremental refinement of the macro state "loss of coolant occurs in fresh water loop", the conclusion of Rule 2. The conclusion of Rule 1, "cooling system is responsible for overheat", is subject to another refinement. The expert looks at the coolant pressure in a similar way to the case of Rule 2 (see Figure 2).

Rule 10:   [1] AND (10) = = = > [10].

      IF   [Cooling system is responsible for overheat]

      AND   (Pressure gauge reading of fresh water is normal)

      THEN [Enough coolant circulates in fresh water loop but engine overheats]

This prerequisite has the same state as Rule 2, but the observable fact differs. This difference leads to the different conclusion. Attention is now focused on cases where there is enough coolant in the fresh water but the engine overheats, the conclusion of Rule 10. The expert first examines the warmup valve, the last component to be checked in the fresh water loop.

Rule 11:   [10] AND (11) = = = > ⟨11⟩.

      IF   [Enough coolant circulates in fresh water loop but engine overheats]

      AND   (Warmup steam valve is open)

      THEN ⟨Failure to close warmup valve after engine warmup is responsible for overheat⟩

The causes associated with the fresh water loop have been analyzed. The expert now moves downstream to find the causes in the heat exchanger or the sea water loop.

Rule 12:   [10] AND (12) = = = > [12].

      IF   [Enough coolant circulates in fresh water loop but engine overheats]

      AND   (Temperature guage reading of fresh water at exit of heat exchanger is high)

      THEN [Heat exchanger is ineffective]

Note that the expert implicitly assumes the event "warm-up steam valve is closed" in the prerequisite of Rule 12. The implicit assumption occurs in the expert mind because Rule 11 is established first and then Rule 12. This problem is closely related to the if-then-else block of structured programming, and is discussed in part A.3. The conclusion of Rule 12 is still a macro state which may have causes in the heat exchanger itself or in the sea water loop. The following rules increase the degree of resolution.

Rule 13:   [12] AND (13) = = = > ⟨13⟩.

      IF   [Heat exchanger is ineffective]

      AND   (Temperature gauge reading of sea water at exit of heat exchanger is low)

      THEN ⟨Fouled heat exchanger and hence poor heat transfer is responsible for overheat⟩

The heat transfer rate is low and the temperature of the sea water does not increase as expected through the heat exchanger. The conclusion of Rule 13 is an ultimate resolution of the plant state and can be a cause of engine overheat. The examiner should next move further downstream, i.e., to the sea water loop.

Rule 14:   [12] AND (14) = = = > [14].
  IF   [Heat exchanger is ineffective]
  AND   (Temperature gauge reading of sea water at exit of heat exchanger is high)
  THEN   [Loss of coolant occurs in sea water loop]

This conclusion can be analyzed analogously to "loss of coolant occurs in fresh water loop", the conclusion of Rule 2. Since the sea water loop has the flow control valve, the expert first looks at the control valve in Rule 15, and then the coolant pump status in Rules 16 and 17, although this is not necessarily an absolute choice.

Rule 15:   [14] AND (15) = = = > ⟨15⟩.
  IF   [Loss of coolant occurs in sea water loop]
  AND   (Sea water control valve is closed)
  THEN   ⟨Sea water control valve closure is responsible for overheat⟩

Although this conclusion may be regarded as an ultimate cause, it can also be decomposed into these causes: (1) the temperature sensor for the fresh water loop is biased low (command failure to the valve), (2) failure of the valve actuator by loss of pneumatic power, etc., (3) the valve is stuck in the closed position (hardware failure of the valve itself).

Rule 16:   [14] AND (16) = = = > ⟨16⟩.
  IF   [Loss of coolant occurs in sea water loop]
  AND   (Sea water coolant pump is not operating)
  THEN   ⟨Sea water coolant pump stoppage is responsible for overheat⟩

Rule 17:   [14] AND (17) = = = > [17].
  IF   [Loss of coolant occurs in sea water loop]
  AND   (Sea water coolant pump is operating)
  THEN   [Loss of coolant occurs in sea water loop with pump operating]

Rule 18:   [17] AND (18) = = = > ⟨18⟩.
  IF   [Loss of coolant occurs in sea water loop with pump operating]
  AND   (Pressure gauge reading of sea water is high)
  THEN   ⟨Sea water discharge valve closure is responsible for overheat⟩

The following four rules are similar to Rules 5, 6, 7, and 9, respectively.

Rule 19:   [17] AND (19) = = = > ⟨19⟩.
  IF   [Loss of coolant occurs in sea water loop with pump operating]
  AND   (Pressure gauge reading at suction valve of sea water coolant pump is low)
  THEN   ⟨Blockage of strainer or closed suction valve of sea water coolant pump is responsible for overheat⟩

Rule 20:   [17] AND (20) = = = > ⟨20⟩.

> **IF** [Loss of coolant occurs in sea water loop with pump operating]
> **AND** (Pressure gauge reading at delivery valve of sea water coolant pump is high)
> **THEN** ⟨Closed delivery valve of sea water coolant pump is responsible for overheat⟩

Rule 21: [17] AND (21) = = = > ⟨21⟩.

> **IF** [Loss of coolant occurs in sea water loop with pump operating]
> **AND** (Ammeter reading of motor of sea water coolant pump is abnormal)
> **THEN** ⟨Power decrease in motor of sea water coolant pump is responsible for overheat⟩

Rule 22: [17] AND (22_1) AND (22_1) = = = > ⟨22⟩.

> **IF** [Loss of coolant occurs in sea water loop with pump operating]
> **AND** (Pressure gauge reading at delivery valve of sea water coolant pump is low)
> **AND** (Pressure gauge reading at suction valve of sea water coolant pump is normal)
> **THEN** ⟨Impeller damage in sea water coolant pump is responsible for overheat⟩

These 22 rules imply 16 causes of the engine overheat associated with the cooling system.

3. Implicit IF-THEN-ELSE Attitude of Expert

A set of observable facts which ensures a plant state is called a cut set of the state, a terminology borrowed from the fault tree analysis. A minimal (min) cut set is such that if a fact is removed from the set, it is no longer a cut set. We can easily identify the minimal cut sets of a plant state, given the AND/OR tree. For instance, consider the AND/OR tree in Figure 2. The conclusion of Rule 12, i.e., "heat exchanger is ineffective" has the min cut set consisting of the following facts: F1 — Engine itself is not the cause, F10 — Pressure gauge reading of fresh water is normal, F12 — Temperature gauge reading of fresh water at exit of heat exchanger is high. On the other hand, the conclusion of Rule 11 has the min cut set consisting of: F1 — Engine itself is not the cause, F10 — Pressure gauge reading of fresh water is normal, and F11 — Warmup steam valve is open.

Fact F10 ensures that there is no line blockage in the fresh water loop. Assume that the warmup steam valve is the single cause of engine overheat. Then, facts F1, F10, and F11 hold, and the cause can be identified successfully. However, fact F12 also holds under this situation because the steam flow would heat up the fresh water loop. Thus, there is also the state, "heat exchanger is ineffective". The sea water temperature at exit of heat exchanger would rise, yielding the conclusion of Rule 14, i.e., "loss of coolant occurs in sea water loop". Since the warmup steam valve is the only cause, none of the observable facts hold in Rules 15 through 22, and the diagnosis happens to finish correctly.

As noted in Part A.2, the expert implicitly assumed in his/her mind the event "warmup steam valve is closed" in the prerequisite of Rule 12. Consequently, Rule 12 lacks the condition and inadvertently allowed its conclusion to hold. In general, such a relaxation of the prerequisite is not so harmful because the fault diagnosis may conclude the true cause together with some spurious causes. We can revise the rule base when the spurious causes are encountered.

The IF-THEN-ELSE structure is useful to avoid the inadvertent relaxation of the prerequisite. For instance, Rules 11 and 12 can be stated simply as:

Rule A:
> **IF**   [Enough coolant circulates in fresh water loop but engine overheats]
> **THEN**
>> **IF**   (Warmup steam valve is open)
>> **THEN** ⟨Failure to close warmup valve after engine warmup is responsible for overheat⟩
>> **ELSEIF**(Temperature gauge reading of fresh water at exit of heat exchanger is high)
>> **THEN**   [Heat exchanger is ineffective] **ENDIF**
>> **ENDIF**
> **ENDIF**

These statements are equivalent to the Rule 11 and a revised version of Rule 12:

Rule 12:
> **IF**   [Enough coolant circulates in fresh water loop but engine overheats]
> **AND**   (Warm-up steam valve is closed)
> **AND**   (Temperature gauge reading of fresh water at exit of heat exchanger is high)
> **THEN**   [Heat exchanger is ineffective]

The AND/OR tree of Figure 2 can be represented as nested if-then-else blocks. This suggests a mixed exploitation of procedural and rule-based programmings.

B. Ship Engine Startup System
1. Startup Operation
Figure 4 is a schematic diagram of a ship engine startup system. Two control levers are observed in the middle of the diagram: main and reverse levers.

1.   **Main lever:** This starts or stops the engine. There are three positions and one analog range arranged in the following sequence.

- STOP: The engine stops at this position.
- AIR-RUN: The engine is run by pressurized air. The air-run feature corresponds to a cell motor of automobiles. This also expels hazardous water remaining in the engine, preventing piston damage at the fuel-run when the engine is running by fuel. At the engine fuel-run, the air-run can be used as a brake.
- AIR-TO-FUEL: The air-run is terminated at this position, and the fuel-run is ready for its start.
- FUEL-RUN: This is an analog range for the engine acceleration by increasing the fuel supply. The engine is accelerated until a steady-state engine speed is attained.

2.   **Reverse lever:** This selects the forward or backward rotations of the screws, and has three positions arranged in the following sequence.

- FORWARD: The screws are directly driven by the engine via crank shafts. The forward rotation results when the engine is coupled with "forward cams" on cam shafts. The FORWARD lever position shifts the cam shafts, implementing the forward cams.
- LATCH: This locks the cam shafts at a forward or a backward position. The screws are now ready for rotation.
- BACKWARD: This shifts the cam shafts to the backward position, implementing

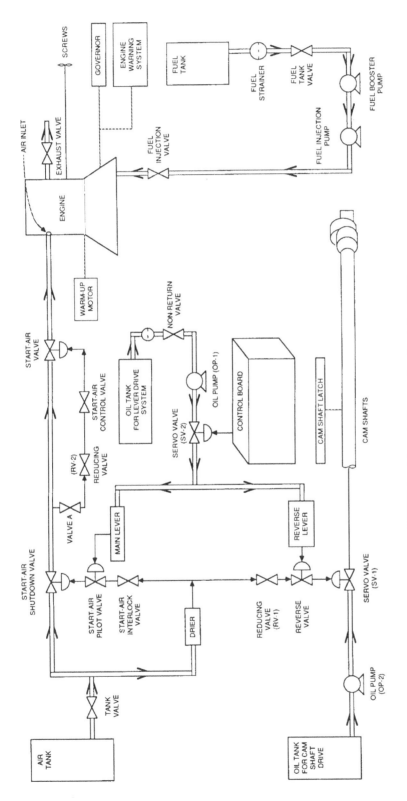

FIGURE 4.   Schematic diagram of engine startup system.

"backward cams". Note that the cam shafts should be located at either forward or backward position. There is no neutral position for the cam shafts.

A warmup motor is shown near the engine in the diagram. This is an electric motor to rotate the engine at the warmup mode (see the warmup valve in Figure 3). The rotation allows the temperature to distribute evenly over the engine. The motor must be detached after the completion of the warmup, otherwise the running engine will destroy the motor and may even injure persons around the motor. The engine startup operation is considered to follow the warmup. The motor may still be coupled with the engine if human error is involved.

The startup operation can be performed from an air-conditioned monitor room near the engine. The main and reverse levers are located in the room. The operation consists of the following steps:

1. Set the reverse lever at the FORWARD or BACKWARD position. This activates the reverse valve in Figure 4 and allows the air to flow in a forward or backward direction which, in turn, activates the servo valve (SV-1) to shift the cam shafts. Forward or backward cams are linked with the engine.
2. Set the reverse lever at the **LATCH** position. This closes the reverse valve which, in turn, deactivates the SV-1 and latches the cam shafts.
3. Set the main lever at the AIR-RUN position. This opens the start-air pilot valve which, in turn, opens the start-air shutdown valve, causing the airflow to the start-air valve. The start-air control valve repeatedly opens and closes the start-air valve, controlling the airflow through inlets to the six-cylinder engine.
4. Set the main lever at the AIR-TO-FUEL position. This closes the start-air pilot valve, closing the start-air shutdown valve, and thus terminating the air-run.
5. Leave the AIR-TO-FUEL position and move the main lever into the FUEL-RUN range to accelerate the engine until a steady engine speed is attained. The fuel gas is supplied to the engine from the fuel tank via strainer, fuel tank valve, fuel booster pump, fuel injection pump, and injection valve.

The start-up operation from the engine monitor room is called an engine-site operation. The engine can also be started up remotely from a control board in the bridge. This is implemented by a remote controller of the main and reverse levers by a pressurized oil system. Two types of operations are available:

1. Remote-automatic: The operation is programmed beforehand, and one may set a navigation dial on the control board. The 12:00 position denotes the main lever STOP, the clockwise range from 12 to 6 represents the FORWARD with increasing engine speed, and the counterclockwise range from 12 to 6 the BACKWARD. The main lever and the reverse levers move automatically in the monitor room by the lever drive oil pressure system.
2. Remote-manual: The startup operation can be performed on the control panel, but more manipulations are required to simulate the engine-site operation.

One must first start the oil pump (OP-1) of the lever drive system. SV-2s control the main and reverse levers which, in turn, start up the engine accordingly. Only the remote-automatic mode is considered in the remainder of Part B.

2. Control, Protection, and Interlocks

The system in Figure 4 has a number of features to maintain its consistency. They include:

1.   A governor on the right edge of Figure 4 controls the engine speed by regulating the fuel flow.
2.   The engine rpm is monitored. The air-run is switched automatically to the fuel-run when the rpm reaches a set point value called an air-cut.
3.   When the engine rpm at the air-run exceeds a value, the main lever cannot move to the AIR-TO-FUEL position, thus prohibiting the fuel-run. Misfires could occur at such a high rpm, resulting in an inadvertent reverse rotation of the screws. The high rpm is often observed when the air-run is used as an emergency brake.
4.   The main and reverse levers are latched when the ship is at anchor. This prevents inadvertent ship propulsions.
5.   The lubrication oil pressures of engine and turbo-charger are monitored (see engine warning system in Figure 4). The engine is shut down when something is wrong with the pressure. Some of the protection systems use a DC power source. The engine is shut down when the DC power is lost. A panic button also initiates the engine shutdown.
6.   As described in Part B.1, a remote control system drives the main and reverse levers at the remote modes. The normal oil pressure for the remote control must be in the range (25 kg/cm², 35 kg/cm²). The OP-1 trips when the pressure deviates from the normal range. It also trips when a timer runs out but the pressure does not reach the normal range.
7.   An electric motor drives the pistons during the engine warmup. Neither air-run nor lever drive OP-1 can start if the motor is left coupled with the engine.
8.   The control board operation of the remote-automatic mode is ineffective if the remote-manual or engine-site mode has been selected. The lever drive OP-1 does not start at the engine-site mode. Reset the startup system and select a suitable mode when the system is recovered from an AC power failure.
9.   The cam shafts must be positioned and latched at a correct location, i.e., FORWARD or BACKWARD, otherwise, the start-air pilot valve does not open, prohibiting the air-run and the lever drive OP-1 does not start.
10.   The OP-2 for cam shaft shift trips by low or high oil pressure.

3. Fuel-Run Failure at Remote-Automatic Mode

Consider an operator who assumes the remote-automatic mode. He must turn the dial on the control board clockwise or counterclockwise to obtain a forward or backward screw rotation at a suitable speed. If everything is all right, the engine starts up automatically in the following way.

1.   The lever drive OP-1 starts. The reverse lever in the monitor room near the engine is set at the FORWARD or BACKWARD position by the oil pressure system. This causes the reverse valve to open which, in turn, activates the cam shaft drive oil pressure system. The cam shafts are started shifting, and are located at the FORWARD or BACKWARD position. The reverse lever then return to the LATCH position, causing the reverse valve to close. This deactivates the shaft drive oil pressure system, and the cam shafts are latched by mechanical stoppers.
2.   The lever drive oil pressure system locates the main lever at the AIR-RUN position, causing the start-air pilot valve to open. The start-air interlock valve also opens when the cam shafts have been latched in the right position. The pilot air then opens the start-air shutdown valve, causing the main air to flow to the start-air valve. The control valve controls the start-air valve which supplies the air to the engine. The air-run starts.
3.   The main lever moves forward to the AIR-TO-FUEL position when the air-run reaches a prespecified engine rpm, i.e., an air cut value. The start-air pilot valve closes,

shutting off the air supply to the engine. The main lever then moves further to a suitable position in the FUEL-RUN range, and activates the fuel supply system. The fuel gas flow rate increases until the engine reaches the rpm specified by the dial on the control board.

The fuel-run would not attained if something is wrong with the startup system. The event which triggers the fault diagnosis is the fuel-run failure. In the next section, the IF-THEN rules which isolate the causes are described.

4. IF-THEN Rules For Isolating Causes Of Fuel-Run Failure
a. Preliminary Refinements
The first step of the incremental refinement is made by the three rules.

Rule 1:

IF	[Fuel-run failure]
AND	(Normal air-run)
THEN	[Fuel-run failure under sufficient air-run rpm]

Rule 2:

IF	[Fuel-run failure]
AND	(Partial air-run)
THEN	[Fuel-run failure by low air-run rpm]

Rule 3:

IF	[Fuel-run failure]
AND	(Complete air-run failure)
THEN	[Fuel-run failure by zero air-run rpm]

The normal air-run denotes that sufficient air-run rpm is reached, but the fuel-run is failed. The partial air-run indicates that the engine starts, but does not reach the sufficient rpm for the fuel-run. The complete air-run failure means that the engine pistons never move.

i. Normal Air-Run
The following three rules refine Rule 1.

Rule 1.1:

IF	[Fuel-run failure under sufficient air-run rpm]
AND	(Main lever at FUEL-RUN position)
THEN	[Fuel-run failure by failed ignition]

Rule 1.2:

IF	[Fuel-run failure under sufficient air-run rpm]
AND	(Main lever at AIR-TO-FUEL position)
THEN	[Fuel-run failure by engine rpm interlock]

Rule 1.3:

IF	[Fuel-run failure under sufficient air-run rpm]
AND	(Main lever at AIR-RUN position)
THEN	[Fuel-run failure by failed lever drive system]

The main lever at the FUEL-RUN position in Rule 1.1 indicates that only ignition failures are causes of the fuel-run failure. The failed ignition can be further refined obtaining a set of ultimate causes of the fuel-run failure.

Rule 1.1.1:

IF	[Fuel-run failure by failed ignition]

> **AND**   (Low cooling water temperature)
> **THEN**   ⟨Insufficient warm-up is responsible for fuel-run failure⟩

Rule 1.1.2:

> **IF**   [Fuel-run failure by failed ignition]
> **AND**   (Fuel tank is closed)
> **THEN**   ⟨Fuel tank main valve closure is responsible for fuel-run failure⟩

Rule 1.1.3:

> **IF**   [Fuel-run failure by failed ignition]
> **AND**   (Low suction pressure of fuel booster pump)
> **AND**   (Fuel tank valve is open)
> **THEN**   ⟨Fuel strainer blockage is responsible for fuel-run failure⟩

Rule 1.1.4:

> **IF**   [Fuel-run failure by failed ignition]
> **AND**   (Low fuel temperature)
> **THEN**   ⟨High fuel viscosity is responsible for fuel-run failure⟩

Rule 1.1.5:

> **IF**   [Fuel-run failure by failed ignition]
> **AND**   (Fuel priming was not performed)
> **THEN**   ⟨Bubbles in fuel injection pipes are responsible for fuel-run failure⟩

Rule 1.1.6:

> **IF**   [Fuel-run failure by failed ignition]
> **AND**   (Low compression pressure in engine cylinders)
> **THEN**   ⟨Loose exhaust valve or defective piston is responsible for fuel-run failure⟩

Rule 1.1.7:

> **IF**   [Fuel-run failure by failed ignition]
> **AND**   (Lack of fuel injection noise)
> **THEN**   ⟨Closed fuel injection valve is responsible for fuel-run failure⟩

Rule 1.1.8:

> **IF**   [Fuel-run failure by failed ignition]
> **AND**   (Abnormal speed indications of governor)
> **THEN**   ⟨Insufficient alignment of governor is responsible for fuel-run failure⟩

The main lever at the AIR-TO-FUEL position in Rule 1.2 suggests that an engine rpm interlock stops the lever at this position, prohibiting the fuel-run. The interlock may operate correctly or spuriously, as shown by the following rules.

Rule 1.2.1:

> **IF**   [Fuel-run failure by engine rpm interlock]
> **AND**   (Air-run speed greater than 50 rpm)
> **THEN**   ⟨Engine rpm interlock operates correctly and stops fuel-run⟩

Rule 1.2.2:

> **IF**   [Fuel-run failure by engine rpm interlock]
> **AND**   (Air-run speed no more than 50 rpm)
> **THEN**   ⟨Engine rpm interlock operates spuriously and stops fuel-run⟩

The main lever at the AIR-RUN position in Rule 1.3 indicates that something is wrong with the remote lever drive system, which is shown by the following rules.

Rule 1.3.1:

> **IF**   [Fuel-run failure by failed lever drive system]

	AND	(Main lever SV-2 operates incorrectly)
	THEN	⟨Main lever SV-2 failure is responsible for fuel-run failure⟩

Rule 1.3.2:

	IF	[Fuel-run failure by failed lever drive system]
	AND	(Main lever drive oil leaks)
	THEN	⟨Oil leakage of main lever drive system is responsible for fuel-run failure⟩

Rule 1.3.3:

	IF	[Fuel-run failure by failed lever drive system]
	AND	(Main lever is latched partially)
	THEN	⟨Partially stuck main lever stopper is responsible for fuel-run failure⟩

The main lever is latched during the anchorage. It must be unlatched at the engine startup. The main lever has moved from STOP to AIR-RUN, indicating that it is partially stuck. These three rules could have been applied to the main lever at the AIR-TO-FUEL position; however, they were neglected in Rule 1.2. The rpm interlock is a more plausible cause because the main lever has moved up to the AIR-TO-FUEL position via AIR-RUN.

ii. Partial Air-Run

More on the conclusion of Rule 2: The air-run is partially possible, and we can assume that the main lever is at the AIR-RUN position. The following refinements hold.

Rule 2.1:

	IF	[Fuel-run failure by low air-run rpm]
	AND	(Remote-manual mode lamp is ON)
	THEN	⟨Remote-manual mode is responsible for fuel-run failure⟩

Rule 2.2:

	IF	[Fuel-run failure by low air-run rpm]
	AND	(Remote-automatic lamp is ON)
	THEN	[Fuel-run failure by low air-run rpm with remote-automatic lamp ON]

The final conclusion can further be refined by the rules that follow.

Rule 2.2.1:

	IF	[Fuel-run failure by low air-run rpm with remote-automatic lamp ON]
	AND	(Air cut value is lower than 40 rpm)
	THEN	⟨Low air cut value is responsible for fuel-run failure⟩

The air cut value is a threshold to terminate the air-run, and thus the low value results in the partial air-run.

Rule 2.2.2:

	IF	[Fuel-run failure by low air-run rpm with remote-automatic lamp ON]
	AND	(Normal air cut value)
	AND	(Air cut relay is ON)
	THEN	⟨Spurious activation of air cut relay is responsible for fuel-run failure⟩

Rule 2.2.3:

	IF	[Fuel-run failure by low air-run rpm with remote-automatic lamp ON]

> **AND**   (Normal air cut value)
> **AND**   (Air cut relay is OFF)
> **THEN**   [Something is wrong with start-up air system]

This conclusion will be developed similarly to Part B.4.b. The partial air-run is caused there by hangover air pressure. Let us now refine the conclusion of Rule 3.

iii. Complete Air-Run Failure
Rule 3.1:

> **IF**   [Fuel-run failure by zero air-run rpm]
> **AND**   (Main lever at AIR-RUN position)
> **THEN**   [Something is wrong with startup air system]

This is the same conclusion as Rule 2.2.3, and is developed in Part B.4.b.

Rule 3.2:

> **IF**   [Fuel-run failure by zero air-run rpm]
> **AND**   (Main lever at STOP position)
> **THEN**   [Something is wrong with main lever drive system]

Main lever positions other than AIR-RUN or STOP are not possible because the lever can move further to the AIR-TO-FUEL position only if the air-run is successful. The conclusion of Rule 3.2 will be developed in Part B.4.c.

b. Something is Wrong with Start-Up Air System
The expert first looks at the start-air shutdown valve to refine the state.

Rule A:

> **IF**   [Something is wrong with start-up air system]
> **AND**   (Start-air shutdown valve is closed)
> **THEN**   [Something is wrong with start-air shutdown valve]

Rule B:

> **IF**   [Something is wrong with start-up air system]
> **AND**   (Start-air shutdown valve is opened)
> **THEN**   [Something is wrong with start-air valve]

Note that the start-air valve locates downstream of the shutdown valve in Figure 4. The expert next looks at the pilot air to refine the conclusion of Rule A.

i. Something is Wrong with Start-Air Shutdown Valve

Rule A.1:

> **IF**   [Something is wrong with start-air shutdown valve]
> **AND**   (Pilot air is being supplied to start-air shutdown valve)
> **THEN**   ⟨Stuck start-air shutdown valve is responsible for fuel-run failure⟩

Rule A.2:

> **IF**   [Something is wrong with start-air shutdown valve]
> **AND**   (Pilot air is not being supplied to start-air shutdown valve)
> **THEN**   [Something is wrong with start pilot air supply system]

The conclusion of Rule A.2 is refined by the start-air pressure and the air supply to the pilot valve.

Rule A.2.1:

IF   [Something is wrong with start pilot air supply system]

AND   (Start-air pressure is zero)

THEN ⟨Closed main valve of air tank is responsible for fuel-run failure⟩

Rule A.2.2.:

IF   [Something is wrong with start pilot air supply system]

AND   (Start-air pressure is normal)

AND   (Start-air is being supplied to pilot valve)

THEN ⟨Start-air pilot valve is stuck closed or is incorrectly linked with main lever⟩

Rule A.2.3:

IF   [Something is wrong with start pilot air supply system]

AND   (Start-air pressure is normal)

AND   (Start-air is not being supplied to pilot valve)

THEN [Start-air interlock valve is active, i.e., is closed]

The conclusion of Rule A.2.3 implies a correct or spurious activation of the interlock valve.

Rule A.2.3.1:

IF   [Start-air interlock valve is active, i.e., is closed]

AND   (Warmup motor is coupled with engine)

THEN ⟨Coupled warm-up motor is responsible for fuel-run failure⟩

Rule A.2.3.2:

IF   [Start-air interlock valve is active, i.e., is closed]

AND   (Warmup motor is detached from engine)

THEN ⟨Spurious activation of start-air interlock valve is responsible for fuel-run failure⟩

ii. Something is Wrong with Start-Air Valve

The conclusion of Rule B can be developed in a similar way to Rule A. Since the start-air shutdown valve is opened, the conclusion is refined by assuming the normal air pressure.

Rule B.1:

IF   [Something is wrong with start-air valve]

AND   (Pilot air is being supplied to start-air valve)

THEN ⟨Stuck start-air valve is responsible for fuel-run failure⟩

Rule B.2:

IF   [Something is wrong with start-air valve]

AND   (Pilot air is not being supplied to start-air valve)

THEN [Something is wrong with start-air control system]

Rule B.2.1:

IF   [Something is wrong with start-air control system]

AND   (Zero air pressure upstream of reducing valve RV-2)

THEN ⟨Closed valve A upstream of reducing valve RV-2 is responsible for fuel-run failure⟩

Rule B.2.2:

IF   [Something is wrong with start-air control system]

AND   (Low air pressure from reducing valve RV-2)

THEN ⟨Valve RV-2 failure is responsible for fuel-run failure⟩

Rule B.2.3:

> **IF**   [Something is wrong with start-air control system]
> **AND**   (Normal air pressure from reducing valve RV-2)
> **THEN** ⟨Start-air control valve failure is responsible for fuel-run failure⟩

c. Something is Wrong with Main Lever Drive System

The expert first looks at the oil pressure of the drive system.

Rule C:

> **IF**   [Something is wrong with main lever drive system]
> **AND**   (Abnormal oil pressure of main lever drive system)
> **THEN** [Main lever drive system failure under abnormal oil pressure]

Rule D:

> **IF**   [Something is wrong with main lever drive system]
> **AND**   (Normal oil pressure of main lever drive system)
> **THEN** [Main lever drive system failure under normal oil pressure]

The conclusion of Rule C is refined by observable facts about the OP-1.

Rule C.1:

> **IF**   [Main lever drive system failure under abnormal oil pressure]
> **AND**   (Oil pump OP-1 never starts)
> **THEN** [OP-1 start failure of main lever drive system]

Rule C.2:

> **IF**   [Main lever drive system failure under abnormal oil pressure
> **AND**   (Oil pump OP-1 starts but stops shortly)
> **THEN** [Interlock trips OP-1 of main lever drive system]

Rule C.3:

> **IF**   [Main lever drive system failure under abnormal oil pressure]
> **AND**   (OP-1 is operating normally)
> **THEN** [Abnormal oil pressure of main lever drive system with normal oil
> pump operation]

The conclusion of Rule D, on the other hand, is refined by the mobility of the cam shafts.

Rule D.1:

> **IF**   [Main lever drive system failure under normal oil pressure]
> **AND**   (Cam shafts stay at incorrect position)
> **THEN** [Main lever drive system failure due to incorrect cam shafts position]

Rule D.2:

> **IF**   [Main lever drive system failure under normal oil pressure]
> **AND**   (Cam shafts stay at correct position)
> **THEN** [Main lever drive system failure under correct cam shaft position]

Assume that the command from the remote-automatic mode is the FORWARD position. Then, the position of the cam shaft at the BACKWARD or in between BACKWARD and FORWARD is incorrect.

i. OP-1 Start Failure of Main Lever Drive System

This is the conclusion of Rule C.1, which has the following rules.

Rule C.1.1:
> **IF**   [OP-1 start failure of main lever drive system]
> **AND**   (Engine-site operation mode)
> **THEN** ⟨Engine-site mode is responsible for fuel-run failure⟩

The OP-1 will not start at the engine-site mode.

Rule C.1.2:
> **IF**   [OP-1 start failure of main lever drive system]
> **AND**   (Warmup motor is coupled with engine)
> **THEN** ⟨Coupled warmup motor is responsible for fuel-run failure⟩

The coupled warmup motor does not allow the oil pump to start.

Rule C.1.3:
> **IF**   [OP-1 start failure of main lever drive system]
> **AND**   (Startup system was not reset after AC power failure)
> **THEN** ⟨Startup system must be reset for successful fuel-run⟩

Rule 3 C.1.4:
> **IF**   [OP-1 start failure of main lever drive system]
> **AND**   (Electric power indicator is OFF for main lever drive oil pump)
> **THEN** ⟨Electric power failure of main lever drive oil pump is responsible for fuel-run failure⟩

Rule C.1.5:
> **IF**   [OP-1 start failure of main lever drive system]
> **AND**   (Electric power indicator is ON for main lever drive oil pump)
> **AND**   (OP-1 motor is smelled burning)
> **THEN** ⟨Burned OP-1 motor coil of main lever drive system is responsible for fuel-run failure⟩

Rule C.1.6:
> **IF**   [OP-1 start failure of main lever drive system]
> **AND**   (Electric power indicator is ON for OP-1
> **AND**   (OP-1 start button is OFF on control board)
> **THEN** ⟨OP-1 start button on control board must be pushed for successful fuel-run⟩

Rule C.1.7:
> **IF**   [OP-1 start failure of main lever drive system]
> **AND**   (Electric power indicator is ON for OP-1)
> **AND**   (OP-1 start button is ON on control board)
> **AND**   (OP-1 starter contacts is OFF)
> **THEN** ⟨OP-1 starter equipment failure is responsible for fuel-run failure⟩

ii. Interlock Trips Oil Pump of Main Lever Drive System

The observable fact of Rule C.2 indicates that some interlocks are active, tripping the oil pump. There are two cases of interlock activation; low oil pressure (Rule C.2.1) and excessive oil pressure (Rule C.2.2)

Rule C.2.1:
> **IF**   [Interlock trips OP-1 of main lever drive system]
> **AND**   (Oil pressure does not reach 25 kg/cm^2)
> **THEN** [OP-1 interlock activation by low oil pressure]

Rule C.2.1.1:
        IF   [OP-1 interlock activation by low pressure]
        AND   (Low level of lever drive system oil tank)
        THEN ⟨Low level of lever drive system oil tank is responsible for fuel-run failure⟩

Rule C.2.1.2:
        IF   [OP-1 interlock activation by low oil pressure]
        AND   (Abnormal noise from OP-1)
        THEN ⟨Bubbles in OP-1 of main lever drive system is responsible for fuel-run failure⟩

Rule C.2.1.3:
        IF   [OP-1 interlock activation by low oil pressure]
        AND   (Oil leaks around OP-1)
        THEN ⟨Oil leakage from main lever drive system is responsible for fuel-run failure⟩

Rule C.2.1.4:
        IF   [OP-1 interlock activation by low oil pressure]
        AND   (Low suction pressure of OP-1)
        THEN ⟨Closure of strainer or valve upstream of OP-1 is responsible for fuel-run failure⟩

The last four rules represent the cases when the interlock operates correctly. The following two rules express spurious interlock activations.

Rule C.2.1.5:
        IF   [OP-1 interlock activation by low oil pressure]
        AND   (Pressure timer operates incorrectly)
        THEN ⟨Pressure timer failure is responsible for fuel-run failure⟩

The oil pressure at 10 sec after the pump startup is monitored by a pressure switch. If the timer fails, then the pressure may be measured at an earlier time point, resulting in a spurious interlock activation.

Rule C.2.1.6:
        IF   [OP-1 interlock activation by low oil pressure]
        AND   (Low-pressure switch operates incorrectly)
        THEN ⟨Defective low-pressure switch is responsible for fuel-run failure⟩

Rule C.2.2:
        IF   [Interlock trips OP-1 of main lever drive system]
        AND   (Oil pressure exceeds 35 kg/cm^2)
        THEN [OP-1 interlock activation by high oil pressure]

Rule C.2.2.1:
        IF   [OP-1 interlock activation by high oil pressure]
        AND   (Line blockage downstream of OP-1)
        THEN ⟨Closed valve or filter of oil line downstream of OP-1 is responsible for fuel-run failure⟩

Rule C.2.2.2:
        IF   [OP-1 interlock activation by high pressure]
        AND   (High-pressure switch operates incorrectly)
        THEN ⟨Defective high pressure switch is responsible for fuel-run failure⟩

iii. Abnormal Oil Pressure with Normal OP-1 Operation

Assume that the warmup motor is coupled with the engine. Then the oil system is depressurized by discharging the oil even if the OP-1 starts. The coupled warmup motor is so dangerous that the three protections are implemented for the startup system.

1. The interlock valve on the start pilot air line is closed, prohibiting the air-run.
2. The OP-1 of the lever drive system does not start, preventing the main or reverse lever movements.
3. The oil system is depressurized once the oil pump starts operating.

The following rules are related to the last protection feature.

Rule C.3.1:

IF	[Abnormal oil pressure of main lever drive system with normal OP-1 operation]
AND	(Warmup motor is coupled with engine)
THEN	⟨Coupled warmup motor is responsible for fuel-run failure⟩

Rule C.3.2:

IF	[Abnormal oil pressure of main lever drive system with normal OP-1 operation]
AND	(Warmup motor is detached from engine)
THEN	⟨Failed-sale failure of warmup motor interlock is responsible for fuel-run failure⟩

iv. Main Lever Drive System Failure Due to Incorrect Cam Shaft Position

The cam shafts normally locate at the FORWARD or BACKWARD position, and latched there when the reverse lever position becomes LATCH. Consider a case when a FORWARD command is generated from the control board during the remote-automatic mode. The reverse lever moves from LATCH to FORWARD, opening the reverse valve and allowing the pilot air to flow into the SV-1 of the cam shaft drive system. The pilot air unlatches the shafts and locates them at the FORWARD position by the oil pressure servo mechanism. The shafts move to the FORWARD position if they were at the BACKWARD position, or they stay at the same position if they were at the FORWARD position. The reverse lever then returns to the LATCH, closes the reverse valve, and latches the cam shafts. The engine is ready for its startup.

The incorrect shaft position occurs if (1) neither FORWARD nor BACKWARD command is being generated from the control board, (2) the reverse lever does not move according to the control board command, (3) the pilot air fails to respond to the reverse lever position (4) the latch mechanism is being stuck, (5) the cam shaft drive system is failed, or (6) some interlocks are being activated to latch the shafts. The expert first looks at the latch status to refine the conclusion of Rule D.1.

Rule D.1.1:

IF	[Main lever drive system failure due to incorrect cam shaft position]
AND	(Cam shafts are unlatched)
THEN	[Cam shafts drive system failure downstream of latch mechanism]

Rule D.1.2:

IF	[Main lever drive system failure due to incorrect cam shaft position]
AND	(Cam shafts remain latched)
THEN	[Cam shafts drive system failure upstream of latch mechanism inclusive]

The observable fact of Rule D.1.1 indicates that the pilot air is being supplied, and hence that the reverse lever must have been moved to the correct position. The conclusion can be refined in the following way.

Rule D.1.1.1:

IF	[Cam shafts drive system failure downstream of latch mechanism]
AND	(Oil leakage from cam shafts drive system)
THEN	⟨Cam shafts drive system oil leakage is responsible for fuel-run failure⟩

Rule D.1.1.2:

IF	[Cam shafts drive system failure downstream of latch mechanism]
AND	(Some valves are inadvertently closed downstream of latch mechanism)
THEN	⟨Line blockage downstream of latch mechanism is responsible for fuel-run failure⟩

Rule D.1.1.3:

IF	[Cam shafts drive system failure downstream of latch mechanism]
AND	(Low levels of oil tanks of cam shafts drive system)
THEN	⟨Oil tank low levels of cam shafts drive system are responsible for fuel-run failure⟩

The latched shafts in Rule D.1.2 yield the following refinement.

Rule D.1.2.1:

IF	[Cam shafts drive system failure upstream of latch mechanism inclusive]
AND	(Cam shafts latch is being stuck)
THEN	⟨Stuck cam shafts latch is responsible for fuel-run failure⟩

Rule D.1.2.2:

IF	[Cam shafts drive system failure upstream of latch mechanism inclusive]
AND	(Engine speed greater than 30 rpm)
THEN	⟨Cam shafts latched by excessive engine speed are responsible for fuel-run failure⟩

This could occur when the FORWARD navigation is suddenly switched to the BACKWARD to avoid collisions. The excessive engine speed may damage the cam shafts, and the interlock latches them.

Rule D.1.2.3:

IF	[Cam shafts drive system failure upstream of latch mechanism inclusive]
AND	(Navigation dial on control board is at STOP position)
THEN	⟨STOP navigation dial is responsible for fuel-run failure⟩

Rule D.1.2.4:

IF	[Cam shafts drive system failure upstream of latch mechanism inclusive]
AND	(Navigation dial on control board is at FORWARD or BACKWARD position)
AND	(Reverse lever stopper is ON)
THEN	⟨Stuck reverse lever stopper is responsible for fuel-run failure⟩

Rule D.1.2.5:

    IF   [Cam shafts drive system failure upstream of latch mechanism inclusive]

    AND (Navigation dial on control board is at FORWARD or BACKWARD position)

    AND (Reverse lever stopper is OFF)

    AND (Reverse lever is at LATCH)

    THEN ⟨Failure of solenoid valve for shifting the reverse lever is responsible for fuel-run failure⟩

Rule D.1.2.6:

    IF   [Cam shafts drive system failure upstream of latch mechanism inclusive]

    AND (Navigation dial on control board is at FORWARD or BACKWARD position)

    AND (Reverse valve is closed)

    THEN ⟨Stuck reverse valve is responsible for fuel-run failure⟩

Rule D.1.2.7:

    IF   [Cam shafts drive system failure upstream of latch mechanism inclusive]

    AND (Reverse valve is open)

    AND (Pilot air pressure through reverse valve is low)

    THEN ⟨Defective pressure reducing valve RV-1 upstream of reverse valve is responsible for fuel-run failure⟩

Rule D.1.2.8:

    IF   [Cam shafts drive system failure upstream of latch mechanism inclusive]

    AND (Normal air pressure through reverse valve)

    AND (Normal engine speed no more than 40 rpm)

    THEN ⟨Spurious engine speed interlock is responsible for fuel-run failure⟩

v. Main Lever Drive System Failure Under Correct Cam Shaft Position

Even if the cam shafts occupy a correct position, the engine does not start if its alarm system is activated. The basic rules follow.

Rule D.2.1:

    IF   [Main lever drive system failure under correct cam shaft position]

    AND (Low lubrication oil pressure of engine)

    THEN ⟨Loss of engine lubricants is responsible for fuel-run failure⟩

Rule D.2.2:

    IF   [Main lever drive system failure under correct cam shaft position]

    AND (Low lubrication oil pressure of turbo-charger)

    THEN ⟨Loss of turbo-charger lubricants is responsible for fuel-run failure⟩

Rule D.2.3:

    IF   [Main lever drive system failure under correct cam shaft position]

    AND (DC 24 V power source failure)

    THEN ⟨DC 24 V power source failure is responsible for fuel-run failure⟩

Rule D.2.4:

    IF   [Main lever drive system failure under correct cam shaft position]

    AND (Engine panic button is ON)

    THEN ⟨Engine panic button activation is responsible for fuel-run failure⟩

Rule D.2.5:

IF	[Main lever drive system failure under correct cam shaft position]
AND	(Enough engine lubricants)
AND	(Enough turbo-charger lubricants)
AND	(DC 24 V ON)
AND	(Engine panic button is OFF)
THEN	⟨Spurious activation of engine warning system is responsible for fuel-run failure⟩

V. A MICRO EXPERT SYSTEM

This section overviews an expert system written in the DML of K-MAN, a PC-based general database management system.

A. Knowledge Base

The knowledge base consists of three types of databases: rule-base, state-base, and fact-base.

1. Rule-Base

This describes the AND/OR tree associated with a set of IF-THEN rules. Each rule follows the syntax of Rule (5). The rule-base has the following fields:

1. RULE-NAME: Name of a rule.
2. PARENT-STATE-NAME: Name of state i.
3. CHILD-STATE-NAME: Name of state j.
4. FACT NAME: Name of an observable fact. There is more than one field of this type. The first field stores the name of fact i_1, and the n-th field the name of fact i_n.
5. JUSTIFICATION: This describes why the rule holds. This is a kind of meta-knowledge for the rule.

Figure 5 shows a rule-base for the AND/OR tree of Figure 2. The JUSTIFICATION field is omitted.

2. State-Base

This describes all the states named in the rule-base. The state-base consists of the following fields:

1. STATE-NAME: Name of a state. This is either a parent state name or a child state name in the rule-base.
2. STATE-DESCRIPTION: The state description appearing in the if-then rule. For instance, State 12 in Rule 12 of Section IV.A.2 has the description, "heat exchanger is ineffective".
3. INFERENCE-POSITION: This describes a degree of incremental refinement at the state. For example, we have at State 12 the description, "It is now clear that there is no cause in the fresh water loop and we proceed to find causes in the heat exchanger or in the sea water loop."
4. RECOVERY-MEASURE: This describes methods of recovering the plant from the state. As an example, consider State 19 in Rule 19 of Section IV.A.2, "blockage of strainer or closed suction valve of sea water coolant pump is responsible for overheat". The recovery measure would be "For the strainer blockage, clean up the strainer with

RULE-NAME	P-STATE	C-STATE	FACT-1	FACT-2
RULE 1	ST 0	ST 1	FT 1	
RULE 2	ST 1	ST 2	FT 2	
RULE 3	ST 2	ST 3	FT 3	
RULE 4	ST 2	ST 4	FT 4	
RULE 5	ST 4	ST 5	FT 5	
RULE 6	ST 4	ST 6	FT 6	
RULE 7	ST 4	ST 7	FT 7	
RULE 8	ST 4	ST 8	FT 8	
RULE 9	ST 4	ST 9	FT 9-1	FT 9-2
FULE 10	ST 1	ST 10	FT 10	
RULE 11	ST 10	ST 11	FT 11	
RULE 12	ST 10	ST 12	FT 12	
RULE 13	ST 12	ST 13	FT 13	
RULE 14	ST 12	ST 14	FT 14	
RULE 15	ST 14	ST 15	FT 15	
RULE 16	ST 14	ST 16	FT 16	
RULE 17	ST 14	ST 17	FT 17	
RULE 18	ST 17	ST 18	FT 18	
RULE 19	ST 17	ST 19	FT 19	
RULE 20	ST 17	ST 20	FT 20	
RULE 21	ST 17	ST 21	FT 21	
RULE 22	ST 17	ST 22	FT 22-1	FT 22-2

FIGURE 5.   An example of a rule base.

the valves on the both sides closed. Be sure to open the valves after the cleanup. Consult with Manual V-100 to open the suction valve". This field is usually filled for the plant state with the ultimate refinement, i.e., for the cause of the event to be diagnosed. These states locate at the leaves of the AND/OR tree, while the event to be diagnosed at the root of the tree. In the goal-oriented inference described in part B, the state of the leaf is called a hypothesis or a goal.

5. PRIORITY: This is an integer to prioritize the validation sequence of hypotheses in the goal-oriented inference. The hypotheses are arranged according to the descending order of the integer values. The integer is defaulted to zero for intermediate state and a nonzero value can establish a sub-goal in the goal-oriented search.

3. Fact-Base

Each fact in the rule-base is described by the following fields:

1. FACT-NAME: Name of an observable fact. This must coincide with a fact name in the rule-base.
2. FACT-DESCRIPTION: The fact description appearing in the IF-THEN rule. For instance, Fact 4 in Rule 4 of Section IV.A.2 has the description, "fresh water coolant pump is operating".
3. CONFIRMATION: This describes methods of confirming the fact. For instance, Fact 4 has the description, "touch the coolant pump to check for its operation".
4. FACT-VARIABLE: This defines a variable for the fact. As an example, consider Fact 2 in Rule 2, i.e., "pressure gauge reading of fresh water is low". This fact associated with it a numeric variable FWPRES which represents the fresh water pressure. Similarly, the fact "fresh water coolant pump is operating" has a logic variable FWPUMP to represent the operation status of the pump.
5. VARIABLE-TYPE: This describes the type of the variable in field FACT-VARIABLE. The type is either NUMERIC or LOGIC.
6. FACT-FUN: This is a logic function of the variable in field FACT-VARIABLE. The

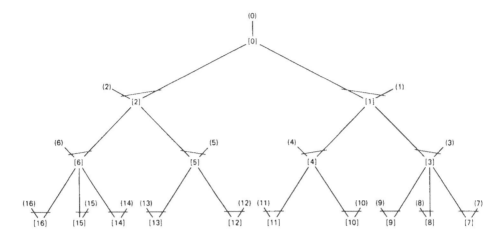

FIGURE 6.   AND/OR tree for description of inference algorithm.

true value means that the fact holds, while the false value rejects the fact. For instance, "pressure gauge reading of fresh water is low" has the function FWPRES $< = 25$ (kg/cm^2). The facts "fresh water coolant pump is operating" and "fresh water coolant pump is not operating" have functions FWPUMP and .NOT. FWPUMP, respectively. The functions stored in the fact-base can be evaluated by a macro statement of K-MAN DML.

7.   COMMENTS: This field stores descriptions which help the user to respond to the questions from the expert system. Such a description typically includes the unit of numeric variable, etc.

B. Inference

Two options are available: data-oriented and goal-oriented inferences. For convenience of description, consider the hypothetical AND/OR tree of Figure 6. The plant states at the botton of the tree are definitive causes of the abnormal event at the top. The top event is assumed to be occurring. Figure 6 indicates ten possible causes or hypotheses [7], [8], ..., [16].

The data-oriented inference is a breadth first search. The expert system first asks the computer operater about the facts (1) and (2). Assume the operator answers that fact (1) holds. The computer then derives state [1] and asks the operator about facts (3) and (4). Assume the operator selects fact (4). The computer derives state [4] and asks the operater about facts (10) and (11). Hypothesis [11] is identified as the cause if the operator selects fact (11).

The goal-oriented inference starts at a hypothesis. Assume that the hypotheses or goals are prioritized according to the sequence [7], [8], ... , [15], [16]. The rightmost cause [7] is first examined. The cause becomes true if state [3] and fact (7) hold. Thus, the goal-oriented inference moves to the examination of the necessary condition [3]. This, in turn, reduces to the confirmation of state [1] and fact (3), and the inference moves toward the necessary condition [1]. This reduces to the confirmation of fact (1) because the top event [0] is known to be occurring.

Fact (1) does not have any prerequisite. Hence, the inference algorithm asks the computer operater whether the fact is true or false. Assume here that the operator answers that the fact is false. Then the algorithm registers fact (1) as the first member of F, a set of false states and facts which are known so far: F = {(1)}. At this point, there are no facts which support state [1] and this state also becomes a member of the set F: F = {(1), [1]}. Moving downward on the tree, note that states [3] and [7] are also rejected: F = {(1), [1], [3], [7]}. Cause [7] is now proven to be false. Moving upward on the AND/OR tree is called "backward chaining" because more basic prerequisites are sought out. Moving downward is called "forward chaining" because this is in the direction from prerequisite to conclusion. The forward or backward chaining is a kind of information retrieval on the AND/OR tree, and can be implemented by the DML.

The goal-oriented inference next begins to confirm cause [8] which has as necessary conditions state [3] and observable fact (8). It is already known that state [3] is false, i.e., [3] is in F. Thus, cause [8] is immediately rejected. In a similar way we can reject causes [9] through [11]. The set F is now given by F = {(1), [1], [3], [7], [8], [9], [10], [11]}.

The inference algorithm now starts confirmation of cause [12]. Moving upward on the tree, the algorithm ultimately reaches observable fact (2). Assume that the operator states that the fact is true. Then the inference algorithm registers fact (2) as the first member of T, a set of true states and facts which are known so far: T = {(2)}. State [2] is automatically true because top state [0] and fact (2) are both true: T = {(2), [2]}. Move downward on the tree to confirm state [5]. Since state [2] is known to be true, the inference algorithm asks the operator about observable fact (5), a necessary condition for state [5]. Assume that the operator replies that the fact is false. Then, the set F is revised to become F = {(1), [1], [3], [7], [8], [9], [10], [11], (5)}. State [5] and hence state [12] are automatically rejected and registered in the set F: F = {(1), [1], [3], [7], [8], [9], [10], [11], (5), [5], [12]}.

Cause [13] is examined in turn. However, state [5], a necessary condition for the cause, is known to be false and cause [13] is rejected: F = {(1), [1], [3], [7], [8], [9], [10], [11], (5), [5], [12], [13]}.

The inference algorithm starts examining cause [14]. This reduces to the confirmation of [6] and (14), and state [6] has two necessary conditions [2] and (6). The first condition is known to be true, i.e., state [2] is in T. The inference algorithm asks the operator about fact (6). Assume that the operator answers "true". This proves state [6] and the set T becomes T = {(2), [2], (6), [6]}. Cause [14] now has a single necessary condition (14) which must be confirmed via the operator. Suppose that fact (14) turns out to be false, rejecting cause [14]: F = {(1), [1], [3], [7], [8], [9], [10], [11], (5), [5], [12], [13], (14), [14]}.

Cause [15] is now inspected for confirmation. In its backward chaining it is found that state [6] has been proven, i.e., [6] is in T. The operator is asked about observable fact (15). Assume that the fact turns out to be true. Cause [15] is proven and the inference algorithm is terminated with T = {(2), [2], (6), [6], (15), [15]}.

In the goal-oriented inference described above, a total of six questions were asked of the computer operator to confirm observable facts (1), (2), (5), (6), (14), and (15). One is not required to examine all 16 observable facts to infer a cause. Once a state is disproven, all descendent states on the AND/OR tree are automatically rejected and added to the set F. This is a characteristic of hierarchical tree such as Figure 6 which represents an icremental refinement of the top event.

The set T is updated when an observable fact turns out to be true. Here, the set consists not only of true observable facts, but also of plant states which can be derived from these true facts by applications of IF-THEN rules. The backward chaining terminates when a true or false state is encountered because moving upward on the tree is a process of examining whether a necessary condition holds or not.

The data or goal-oriented inferences described above are a "YES-NO" version. As seen from the fact-base structure, the operator can answer by numeric values for a fact, and the inference module evaluates the logic function about the fact. Consider fact 2, "pressure gauge reading of fresh water is low" described by the logic function FWPRES < = 25 (kg/cm^2). The fact holds if the operator gives the pressure value no more than 25. Consider also fact 10 of Rule 10 of Section IV.A.2, "pressure gauge reading of fresh water is normal", described by the function 25 < FWPRES < 35. Then this fact is automatically rejected when fact 2 is confirmed. This is called a "fact correlation".

The numeric value input and the fact correlation seem more intelligent than the YES-NO version. However, the YES-NO version allows the computer operator to examine every step of inference, and thus increases the understandability of the AND/OR tree. The more intelligent inference would be useful after the operator gets used to the knowledge base or the knowledge base revisions are finished.

C. Explanation

Explanation of the fault diagnosis process is one of the positive features specific to the expert system. The current version has the following options: (1) MYCIN like WHY and HOW, (2) confirmation method for an observable fact, (3) inference position at a plant state, (4) plant recovery measure, (5) justification of a rule, (6) a set of hypotheses under a plant state, (7) sets of observable facts for proving a plant state, (8) a set of hypotheses rejected so far, (9) inference history, etc.

D. Database Retrieval

The knowledge base and other databases can be accessed at any stage of fault diagnosis. For instance, rules referring to valves and pumps can be retrieved. The database can include manuals, documents, and personnel information, etc.

E. Database Edit and Debug

Databases can be modified by the K-Man edit commands or procedures. Three databases of the knowledge base are linked, yielding rules in a natural language format. State names and others are checked for their consistency. The knowledge base can be renamed, deleted, or copied.

VI. APPLICATION TO SHIP ENGINE COOLING SYSTEM

We now apply the ES to the ship engine cooling system described in Section IV.A. The knowledge base consisted of 22 rules and 15 hypotheses. The AND/OR tree was shown in Figure 2. The top event to be diagnosed was the engine overheat.

An inference process is briefly summarized below. The knowledge base for the cooling system is selected, and the data-oriented inference of an YES-NO version is used. The computer operator responses follow symbol "?".

- Engine overheats? TRUE
- Engine itself is not the cause? TRUE
- Rule 1 [Engine overheats] AND (Engine itself is not the cause) deduces that [Cooling system is responsible for overheat]
- Pressure gauge reading of fresh water is low? WHY
- We know that:
    [Cooling system is responsible for overheat]
    If you inform me further that:
    (Pressure gauge reading of fresh water is low)
    Then we will conclude by Rule 2 that:
    [Loss of coolant occurs in fresh water loop]
- Pressure gauge reading of fresh water is low? FALSE
- Pressure gauge reading of fresh water is normal? TRUE
- Rule 10 [Cooling system is responsible for overheat] AND (Pressure gauge reading of fresh water is normal) deduces that [Enough coolant circulates in fresh water loop]

- Warm-up steam valve is open? FALSE
- Temperature gauge reading of fresh water at exit of heat exchanger is high? POSITION
- It is now clear that there is no cause in the fresh water loop and we proceed to find causes in the heat exchanger or in the sea water loop.
- Temperature gauge reading of fresh water at exit of heat exchanger is high? TRUE
- Rule 12 [Enough coolant circulates in fresh water loop] AND (Temperature gauge reading of fresh water at exit of heat exchanger is high) deduces that [Heat exchanger is ineffective]
- Temperature gauge reading of sea water at exit of heat exchanger is low? FALSE
- Temperature gauge reading of sea water at exit of heat exchanger is high? TRUE
- Rule 14 [Heat exchanger is ineffective] AND (Temperature gauge reading of sea water at exit of heat exchanger is high) deduces that [Loss of coolant occurs in sea water loop]
- Sea water control valve is closed? FALSE
- Sea water coolant pump is not operating? CONFIRMATION
- Touch the coolant pump to check for its operation
- Sea water coolant pump is not operating? FALSE
- Sea water coolant pump is operating? TRUE
- Rule 17 [Loss of coolant occurs in sea water loop] AND (Sea water coolant pump is operating) deduces that [Loss of coolant occurs in sea water loop with pump operating]
- Pressure gauge reading of sea water is high? FALSE
- Pressure gauge reading at suction valve of sea water coolant pump is low? TRUE
- Rule 19 [Loss of coolant occurs in sea water loop with pump operating] AND (Pressure gauge reading at suction valve of sea water coolant pump is low) deduces that ⟨Blockage of strainer or closed suction valve of sea water coolant pump is responsible for overheat⟩
- Cause: ⟨Blockage of strainer or closed suction valve of sea water coolant pump is responsible for overheat⟩

VII. CONCLUSION

The expert system for the fault diagnosis is an element of a plant information system, and hence can be implemented by a data manipulation language. The emphasis on expert knowledge is specific to the expert system. Hopefully, the rule bases and system schematics presented in Section IV contribute to further clarification of knowledge acquisition and improvement of expert system architecture.

An expert system language called XL was recently released from ISR (Intelligent Systems Research) in Australia; X-TRACT is an interpreter implementation of XL.[19] Procedural-based, frame-based, rule-based, and event-based programming paradigms are mixed in a world of a three-value logic; true, false, and unknown. This language includes elegant extensions of DML concepts into artificial intelligence fields.

REFERENCES

1. **Roth, F. H., Waterman, D. A., Lenat, D. B., Ed.,** *Building Expert Systems,* Addison-Wesley, Reading, Mass., 1983.
2. **Barr, A. and Feigenbaum, E. A., Ed.,** *The Handbook of Artificial Intelligence,* Vol. 1—3; Pitman Books, London, 1981.
3. **Nilsson, N. J.,** *Principles of Artificial Intelligence,* Springer-Verlag, Berlin, 1980.

4. **Winston, P. H. and Horn, B. K. P.**, *Lisp;* Addison-Wesley, Reading, Mass., 1981.

5. **Clark, K. L. and McCabe, F. G.**, *micro-PROLOG, Programming In Logic,* Prentice-Hall, Englewood Cliffs, N. J., 1984.

6. **Hammond, P. and Sergot, M.**, *APES — Augumented Prolog For Expert Systems,* Logic Based Systems, Ltd., Surrey, U. K., 1984.

7. **Date, C. J.**, *An Introduction To Database Systems.* Addison-Wesley, Reading, Mass., 1975.

8. **Codd, E. F.**, Recent Investigations In Relational Data Base Systems; *Proc. Int. Fed. Inf. Proc.,* (IFIP 74,) 1974, 1017.

9. dBASE II User's Manual 1, Ashton-Tate, Culver City, Calif., 1981.

10. The dBASE II explosion; *PC Magazine,* 3,(2) 1984.

11. Knowledge Man — The Knowledge Manager Reference Manual 1, Micro Data Base Systems, Inc., Lafayette, Ind., 1983.

12. On-Line Power Plant Alarm and Disturbance Analysis, EPRI, NP-1379, Electric Power Research Institute, Palo Alto, Calif., 1980.

13. **Kehler, T. R.**, Industrial strength knowledge bases — issues and experiences, *Proc. IJCAI,* p. 108, 1983.

14. **Kumamoto, H., Ikenishi, K., Inoue, K., and Henley, E. J.**, Application of expert system techniques to fault diagnosis, *Chem. Eng. J. Biol. Eng. J.,* 29(1), 1, 1984.

15. **Andow, P. K.**, Expert systems in process plant fault diagnosis, *Inst. Chem. Eng. Symp. Ser.,* 90, 1984.

16. **Vesonder, G. T. et al.**, ACE — An expert system for telephone cable maintenance; *Proc. IJCAI,* 110, 1983.

17. **Henley, E. J. and Kumamoto, H.**, *Designing For Reliability and Safety Control;* Prentice-Hall, Englewood Cliffs, N.J., 1985.

18. **Henley, E. J. and Kumamoto, H.**, *Reliability Engineering and Risk Assessment,* Prentice-Hall, Englewood Cliffs, N.J., 1985.

19. X-TRACT System Reference Manual, I.S.R., (2nd Floor, 969 Burke Road, Camberwell 3124, Victoria, Australia), 1987.

Chapter 6

FUZZY FAULT TREE ANALYSIS: THEORY AND APPLICATION

F. S. Lai, Sujeet Shenoi, and L. T. Fan

TABLE OF CONTENTS

I. INTRODUCTION

Fault tree analysis[1-5] is a powerful tool for assessing the reliability of complex large-scale systems. A fault tree provides a logical and hierarchical description of an accident (top event) in terms of sequences and combinations of malfunctions of individual components and adverse operating conditions (basic or fundamental events). By resorting to a fault tree, the reliability of a complex system can be computed in terms of the probabilities of occurrence of the basic events.

In conventional fault tree analysis, probabilities are assigned to the basic events of the system under consideration. These probabilities are combined according to the structure of the AND/OR fault tree.

In many instances, it is difficult to estimate exactly the failure rates of individual components or the probabilities of occurrence of undesirable events. Such situations are likely to arise in dynamically changing environments or in systems for which the available data are insufficient for the statistical estimation of probabilities. Often there are extremely hazardous accidents that may have never occurred before, or occur so infrequently that reasonable data are not available.

In the absence of genuine probability data, it may be necessary to incorporate rough estimates of probabilities that may be supplied by systems designers and operations personnel. When such estimates are provided, it is inappropriate to employ conventional fault tree analyses in quantitative risk assessment studies. Morever, the personnel familiar with the system invariably express their knowledge in qualitative and highly subjective terms; they find it extremely difficult to specify the exact numerical values that must be assigned in place of the missing probabilities. Clearly, a formalism is required to capture and engage the subjective notions of probability in fault tree analysis.

Fuzzy set theory[6-8] provides a scheme for coping with the imprecision and ambiguity that arise out of human expression. By resorting to fuzzy set theory, it may be appropriate to employ the concept of "fuzzy probability", viz., a fuzzy set defined on a probability space, in expressing subjective notions of failure probabilities. This idea, in conjunction with an AND/OR fault tree, forms the basis of the fuzzy fault tree model.[9,10]

II. FAULT TREE ANALYSIS

Prior to entering into the details of the fuzzy fault tree model, it is instructive to review conventional fault tree analysis. The first step involves the construction of a fault tree. Naturally, this is also a required step in fuzzy fault tree analysis.

A fault tree[1,3,4] is a logical and hierarchical model of an accident (top event) expressed in terms of all possible sequences and combinations of fundamental events leading to the top event (Figure 1). There are three fundamental types of events in a fault tree model of an accident.

1. An event that corresponds to a primary failure in the system; this is represented by a circle in the fault tree diagram.
2. An event corresponding to a nonprimary failure that is not decomposed into more basic events; this is represented by a diamond in the fault tree.
3. An event that does not correspond to a fault or a failure but is an ordinary event existing inherently within the system. Such an event is usually represented by a pentagon.

Two types of logical operators or gates are available for connecting the various events in a fault tree.

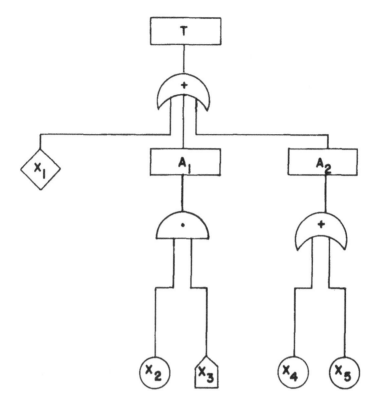

FIGURE 1.   An example of a fault tree.

1. An AND gate links two or more events that must occur simultaneously. The output is an intermediate or top event, usually represented by a rectangle. The symbol ".". stands for an AND operation.
2. An OR gate links two or more events whose independent occurrence is sufficient to result in a specific intermediate or top event (a rectangle in the fault tree). The symbol " + " stands for an OR operation.

The construction of the fault tree is a major task in hazard analysis. It is of immense importance to exhaustively analyze the system prior to constructing the fault tree. This importance cannot be overstressed; in spite of all the precautions usually taken, the majority of errors in hazard analyses of real systems invariably arise out of structural errors in fault trees such as incorrect cause-effect relationships and missing events, rather than errors in the precision of probability data. These errors can lead to potentially disastrous consequences. Safety can only be bought at a price, and it is worthwhile to devote considerable effort in ensuring that all hazard sources are identified.

Although checklists are often employed in industry for the identification of latent hazards, some doubts exist concerning their utility. An inherent disadvantage of the procedure is that many hazards may not surface and may be left out entirely. Perhaps the best method of hazard identification is to provide an opportunity for design and operations personnel to let their imaginations run wild and think of all possible ways in which undesirable situations could arise. However, if the brainstorming is not done in an orderly fashion, a substantial number of hazards may be overlooked.

An operability study[11-14] is an orderly procedure for detecting all conceivable abnormalities within a system. In this study, each subsystem is examined and evaluated in a systematic but creative manner by a team of design and operations personnel. All operational problems

Table 1
LIST OF GUIDE WORDS EMPLOYED IN AN
OPERABILITY STUDY

Guide word	Meaning	Comment
NONE	The complete reverse of the design or operating intentions	No part of the intentions is achieved, but nothing else happens
MORE OF LESS OF	Quantitative increases or decreases	These refer to relevant physical properties such as temperatures and flows
AS WELL AS	A qualitative increase	All the design and operating intentions are achieved together with some additional activities
PART OF	A qualitative decrease	Only some of the intentions are achieved; some are not
REVERSE	The logical opposite of an intention	Mostly applicable to activities
OTHER THAN	Complete substitution	The original intentions are not achieved; something quite different happens

are considered as arising only from deviations from design and operating intentions. The deviations are generated by applying a carefully chosen checklist of guide words to each integral part of the system (Table 1). Questions concerning the possibility, causes, consequences, and modes of prevention of each process deviation are raised. These questions serve to identify and clarify the relationships between the causes and effects of potentially hazardous events and consequently, have an important role in constructing the fault tree.

A sample fault tree is illustrated in Figure 1. The tree describes the occurrence of the top event, T, in terms of the occurrence of five fundamental events X_i, $i = 1, 2, ..., 5$. In some reliability studies, it may be necessary to consider the nonoccurrence of the top event. A so-called dual tree must be constructed. This tree expresses the nonoccurrence of the top event in terms of the nonoccurrence of the fundamental events. To obtain a dual fault tree from an original tree, each OR gate must be replaced by an AND gate, and vice versa. The dual tree corresponding to our sample fault tree is presented in Figure 2.

Having constructed the fault tree, the probability of occurrence of the top event can be easily expressed in terms of the probabilities of the fundamental events. For the tree in Figure 1, the top event, T, may be expressed as

$$T = X_1 \cup A_1 \cup A_2 \tag{1a}$$

where

$$A_1 = X_2 \cap X_3 \tag{1b}$$

and

$$A_2 = X_4 \cup X_5. \tag{1c}$$

Given the probabilities of occurrence of X_i, P_{X_i} ($i = 1, 2, ..., 5$), the top event probability can be written as

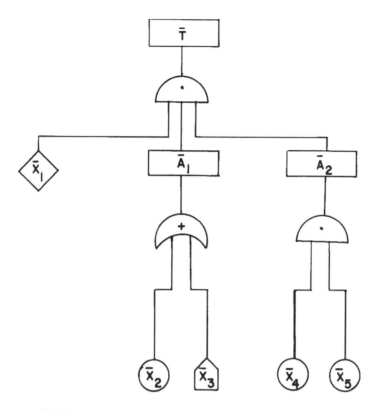

FIGURE 2. The dual tree corresponding to the fault tree in Figure 1.

$$P_T = 1 - (1 - P_{X_1})(1 - P_{A_1})(1 - P_{A_2}) \qquad (2a)$$

where

$$P_{A_1} = P_{X_2} \cdot P_{X_3} \qquad (2b)$$

and

$$P_{A_2} = 1 - (1 - P_{X_4})(1 - P_{X_5}). \qquad (2c)$$

In general, if the event A_m is the output of an AND gate with inputs $A_1, A_2, ..., A_n$, then

$$A_m = A_1 \cap A_2 \cap ... \cap A_n \qquad (3a)$$

and

$$P_{A_m} = P_{A_1} \cdot P_{A_2} \cdot ... \cdot P_{A_n}. \qquad (3b)$$

If event A_m is the output of an OR gate with inputs $A_1, A_2, ..., A_n$, then

$$A_m = A_1 \cup A_2 \cup ... \cup A_n \qquad (4a)$$

and

Table 2
MINIMAL CUT AND
MINIMAL PATH SETS

Minimal cut sets	Minimal path sets
X_1	X_1, X_2, X_4, X_5
X_2, X_3	X_1, X_3, X_4, X_5
X_4	
X_5	

$$P_{A_m} = 1 - (1 - P_{A_1})(1 - P_{A_2}) \dots (1 - P_{A_n}). \tag{4b}$$

Equations 3 and 4 indicate that the probability of the output event, P_{A_m}, is an increasing function of the input event probabilities, P_{A_i} ($i = 1, 2, \dots, n$).

It is easy to show that the probability of occurrence of the top event, P_T, is also an increasing function of the probabilities of occurrence of the basic or fundamental events. This relationship can be employed in determining suitable approaches for increasing the system reliability. For example, if we define

$$P_T \triangleq P_T(P_{X_1}, P_{X_2}, \dots, P_{X_{i-1}}, P_{X_i}, P_{X_{i+x}}, \dots, P_{X_n}) \tag{5a}$$

and

$$P_{T_i} \triangleq P_T(P_{X_1}, P_{X_2}, \dots, P_{X_{i-1}}, 0, P_{X_{i+1}}, \dots, P_{X_n}) \tag{5b}$$

where P_T is probability of the top event, and P_{T_i} is the corresponding probability when the fundamental event X_i has been eliminated. An appropriate index of improvement that measures the difference between P_T and P_{T_i} may be defined:

$$V(P_T, P_{T_i}) = P_T - P_{T_i} > 0. \tag{6}$$

If

$$V(P_T, P_{T_i}) > V(P_T, P_{T_j}), \tag{7}$$

then preventing the occurrence of event X_i is more effective than eliminating event X_j. The index of improvement is a useful measure in system analysis; it may be employed in ranking basic events according to their effectiveness in improving the reliability of the system.

A cut set is a sequence of events that lead to the occurrence of the top event. When the basic events in a cut set cannot be further reduced, it is known as a minimal cut set. The minimum cut sets for the fault tree in Figure 1 are given in Table 2. The listing of minimal cut sets is useful in risk assessment studies; a simple algorithm for determining minimal cut sets has been developed by Fussell and Velsely.[15] Another important concept is the minimal path set. A path set is a collection of basic events whose nonoccurrence ensures the nonoccurrence of the top event. A path set is minimal if it cannot be further reduced and remains a path. Table 2 lists the minimal path sets corresponding to the fault tree in Figure 1. It is interesting to note that the minimal path sets for a given tree can be obtained by employing a minimal cut set algorithm on the dual tree.

III. FUZZY FAULT TREE ANALYSIS

In many real-world applications, it may be difficult to assign exact values to the probabilities of occurrence of the fundamental events. This problem is likely to arise in dynamically changing environments or in systems in which accidents occur so infrequently that reasonable failure data are not available. In the absence of genuine probability data, estimates of failure probabilities are customarily supplied by personnel familiar with the operation of the system. Usually they prefer to express their knowledge in general terms and find it extremely difficult to specify the exact numerical values that are required in conventional fault tree analysis.

To cope with the problems associated with the assignment of exact numerical values to failure probabilities, it is appropriate to employ the concept of "fuzzy probability"[9,10] in representing and articulating human subjective notions of probability. Fuzzy probability can be viewed as a fuzzy set defined on a probability space that expresses the subjective notion that the probability of occurrence of an event is approximately equal to a certain value. For example, when asked for an estimate, an expert plant operator might not be willing to state that the probability of an undesirable event is exactly equal to 0.5; instead, he would definitely prefer to volunteer something of the sort "the probability of occurrence is around 0.5, it is, perhaps, as high as 0.85, and as low as 0.25". A linguistic idea such as this is best expressed by a fuzzy set.[6]

A fuzzy set $\tilde{A}$ is characterized by its membership function, $\mu_{\tilde{A}}$, associated with a value from the interval [0.1] for each element x_j in $\tilde{A}$. The value of the membership function, $\mu_{\tilde{A}}(x_i)$, expresses the degree to which the element x_i belongs to fuzzy set $\tilde{A}$. Fuzzy probability attempts to capture the notion that the value assigned to a probability is a "fuzzy number" between 0 and 1; this fuzzy number is expressed as a fuzzy set. For example, a representation of the number "around 0.5" is illustrated in Figure 3A.

Aside from unimodality, no restriction is imposed on the shape of a fuzzy set that may be employed. However, an important criterion in the present application is the ease of manipulation. The fuzzy probabilities of basic events must be combined in order to compute the fuzzy probability of the top event. This requires an extension of traditional multiplication for fuzzy sets. The resultant operation is rather complex. Consequently, to facilitate computation, the scope of the present work is restricted to the consideration of trapezoidal fuzzy sets. Thus, the fuzzy probability of occurrence of an event X_i is denoted by the four-tuple[9,10]

$$\tilde{P}_{X_i} \triangleq (q_i^{\ell}, p_i^{\ell}, p_i^{r}, q_i^{r}), \tag{8}$$

corresponding to the following membership function

$$\mu_{\tilde{P}_{X_i}}(p) = \begin{cases} 0; & 0 \le p \le q_i^{\ell} \\ 1 - \dfrac{(p_i^{\ell} - p)}{(p_i^{\ell} - q_i^{\ell})}; & q_i^{\ell} \le p \le p_i^{\ell} \\ 1; & p_i^{\ell} \le p \le p_i^{r} \\ 1 - \dfrac{(p - p_i^{r})}{(q_i^{r} - p_i^{r})}; & p_i^{r} \le p \le q_i^{r} \\ 0; & q_i^{r} \le p \le 1. \end{cases} \tag{9}$$

The trapezoidal representation of the concept "around 0.5" is given in Figure 3B.

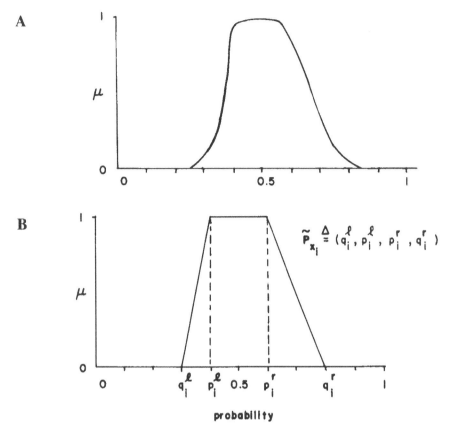

FIGURE 3. (A) Representation of the fuzzy concept "around 0.5". (B) Trapezoidal representation of fuzzy probability "around 0.5".

Since probabilities are expressed in terms of fuzzy sets instead of simple numerical values, it is necessary to employ the fuzzy set counterparts of multiplication ($P_{X_i} \cdot P_{X_j}$) and complementation ($1 - P_{X_i}$) in order to compute the top event probability. The extension principle[8,16] may be employed for this task. It is one of the most important tools in the theory of fuzzy sets and provides a general methodology for "extending" operations in classical mathematics to their equivalents in the domain of fuzzy sets. More specifically, if a certain relationship exists between nonfuzzy entities, it is possible to translate this relationship into one between fuzzy variables by resorting to the extension principle.

We shall not examine the details of the extension principle in this chapter. Nevertheless, through the use of this principle and some simplifications, the multiplication operator ($\odot$)[9,10] defined later can be shown to be a reasonable approximation of true multiplication ($\cdot$) as defined by the extension principle for trapezoidal fuzzy sets.

$$\tilde{P}_{X_i} \odot \tilde{P}_{X_j} \triangleq (q_i^\ell q_j^\ell,\ p_i^\ell p_j^\ell,\ p_i^r p_j^r,\ q_i^r q_j^r)$$

$$\cong \tilde{P}_{X_i} \cdot \tilde{P}_{X_j}. \tag{10}$$

In fact, Tanaka et al.[10] have demonstrated that the following relation holds

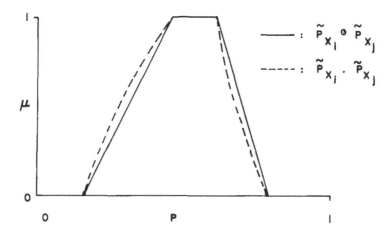

FIGURE 4. Comparison of multiplication operators.

$$\tilde{P}_{X_i} \odot \tilde{P}_{X_j} \gtrsim \tilde{P}_{X_i} \cdot \tilde{P}_{X_j}.$$

This implies that the approximate produce ($\odot$) overestimates the fuzzy probability of failure, thereby providing a greater margin of safety. In any case, as demonstrated in Figure 4, the effect of the approximation is small. The complement of a fuzzy probability is given by

$$1 - \tilde{P}_{X_i} \triangleq (1 - q_i^r, 1 - p_i^r, 1 - p_i^\ell, 1 - q_i^\ell). \tag{11}$$

No approximation is involved in the definition of the complement.

Fuzzy probabilities can be propagated up the fault tree by employing the approximate product and complementation operators. If an event A_m is the output of an AND gate with inputs $A_1, A_2, \ldots, A_n$, then the fuzzy probability is given by

$$\tilde{P}_{A_m} = \tilde{P}_{A_1} \odot \tilde{P}_{A_2} \odot \ldots \odot \tilde{P}_{A_n}. \tag{12}$$

The corresponding fuzzy probability for an OR gate with n input events is expressed as

$$\tilde{P}_{A_m} = 1 - (1 - \tilde{P}_{A_1}) \odot (1 - \tilde{P}_{A_2}) \odot \ldots \odot (1 - \tilde{P}_{A_n}). \tag{13}$$

It is important to note that the logical equation of a fault tree should not be expanded under the approximate product operator. The effects of the approximation are magnified, and the resulting fuzzy probability may be considerably different from that obtained directly from the extension principle. Essentially, each expansion under the approximate product tends to overestimate the fuzzy probability or

$$(1 - \tilde{P}_A) \odot (1 - \tilde{P}_B) \subset 1 - (\tilde{P}_A + \tilde{P}_B) + \tilde{P}_A \odot \tilde{P}_B. \tag{14}$$

If a number of expansions are performed (such a situation could arise in the analysis of large fault trees or when computations are performed with minimal path and cut sets), an excessive margin of safety could be realized. This result is a consequence of a theorem due to Tanaka et al.[10]

Similar to the case involving classical probabilities, the fuzzy probability of occurrence of the top event, $\tilde{P}_T$, is an increasing function (in the fuzzy sense) of the fuzzy probabilities of the basic events. By resorting to this relation, the relative dominance of events may be compared. An appropriate index of improvement[10] in fuzzy fault tree analysis is given by

$$\tilde{V}(\tilde{P}_T, \tilde{P}_{T_i}) = (q_T^\ell - q_{T_i}^\ell) + (p_T^\ell - p_{T_i}^\ell) + (p_T^r - p_{T_i}^r)$$
$$+ (q_T^r - q_{T_i}^r)$$
$$> 0 \tag{15a}$$

$$\tilde{P}_T \triangleq \tilde{P}_T(\tilde{P}_{X_1}, \tilde{P}_{X_2}, \ldots, \tilde{P}_{X_{i-1}}, \tilde{P}_{X_i}, \tilde{P}_{X_{i+1}}, \ldots, \tilde{P}_{X_n})$$
$$\triangleq (q_T^\ell, p_T^\ell, p_T^r, q_T^r) \tag{15b}$$

and

$$\tilde{P}_{T_i} \triangleq \tilde{P}_T(\tilde{P}_{X_1}, \tilde{P}_{X_2}, \ldots, \tilde{P}_{X_{i-1}}, 0, \tilde{P}_{X_{i+1}}, \ldots, \tilde{P}_{X_n})$$
$$\triangleq (q_{T_i}^\ell, p_{T_i}^\ell, p_{T_i}^r, q_{T_i}^r). \tag{15c}$$

The improvement index may be employed in ranking basic events according to their effectiveness in improving system reliability.

IV. ILLUSTRATIVE EXAMPLE

Consider the fault-tree presented in Figure 1. Let us suppose that the personnel familiar with the plant have provided the following subjective assessments of the failure probabilities of the five fundamental events:

$$\tilde{P}_{X_1} \equiv \text{``around } 0.10\text{''}$$

$$\tilde{P}_{X_2} \equiv \text{``around } 0.20\text{''}$$

$$\tilde{P}_{X_3} \equiv \text{``around } 0.30\text{''}$$

$$\tilde{P}_{X_4} \equiv \text{``around } 0.25\text{''}$$

and

$$\tilde{P}_{X_5} \equiv \text{``around } 0.15\text{''}.$$

The subjective assessments of failure probabilities can be expressed by the following trapezoidal fuzzy sets.

$$\tilde{P}_{X_1} \triangleq (0.05, 0.07, 0.12, 0.15)$$

$$\tilde{P}_{X_2} \triangleq (0.12, 0.15, 0.25, 0.27)$$

$$\tilde{P}_{X_3} \triangleq (0.25, 0.28, 0.33, 0.36)$$

$$\tilde{P}_{X_4} \triangleq (0.20, 0.22, 0.27, 0.30)$$

and

$$\tilde{P}_{X_5} \triangleq (0.15, 0.10, 0.20, 0.25).$$

The fuzzy probability of occurrence of the top event is expressed by

$$\tilde{P}_T = 1 - (1 - \tilde{P}_{X_1}) \odot (1 - \tilde{P}_{A_1}) \odot (1 - \tilde{P}_{A_2})$$

where

$$\tilde{P}_{A_1} = \tilde{P}_{X_2} \odot \tilde{P}_{X_3}$$

and

$$\tilde{P}_{A_2} = 1 - (1 - \tilde{P}_{X_4}) \odot (1 - \tilde{P}_{X_5}).$$

Hence, we obtain

$$
\begin{aligned}
\tilde{P}_{A_1} = \tilde{P}_{X_2} \odot \tilde{P}_{X_3} \\
&= (0.12 \times 0.25, 0.15 \times 0.28, 0.25 \times 0.33, 0.27 \times 0.36) \\
&= (0.03, 0.04, 0.08, 0.10)
\end{aligned}
$$

$$
\begin{aligned}
\tilde{P}_{X_{4c}} &= 1 - \tilde{P}_{X_4} \\
&= (1 - 0.30, 1 - 0.27, 1 - 0.22, 1 - 0.20) \\
&= (0.70, 0.73, 0.78, 0.80)
\end{aligned}
$$

$$
\begin{aligned}
\tilde{P}_{X_{5c}} &= 1 - \tilde{P}_{X_5} \\
&= (1 - 0.25, 1 - 0.20, 1 - 0.10, 1 - 0.05) \\
&= (0.75, 0.80, 0.90, 0.95)
\end{aligned}
$$

$$
\begin{aligned}
\tilde{P}_{A_2} &= 1 - \tilde{P}_{X_{4c}} \odot \tilde{P}_{X_{5c}} \\
&= (1 - 0.80 \times 0.95, 1 - 0.78 \times 0.90, \ 1 - 0.73 \times 0.80, \\
&\qquad\qquad\qquad\qquad\qquad\qquad 1 - 0.70 \times 0.75) \\
&= (1 - 0.76, 1 - 0.70, 1 - 0.58, 1 - 0.53)
\end{aligned}
$$

$$= (0.24, 0.30, 0.42, 0.47)$$

$$\tilde{P}_{X_{1c}} = 1 - \tilde{P}_{X_1}$$

$$= (1 - 0.15, 1 - 0.12, 1 - 0.07, 1 - 0.05)$$

$$= (0.85, 0.88, 0.93, 0.95)$$

$$\tilde{P}_{A_{1c}} = 1 - \tilde{P}_{A_1}$$

$$= (1 - 0.10, 1 - 0.08, 1 - 0.04, 1 - 0.03)$$

$$= (0.90, 0.92, 0.96, 0.97)$$

$$\tilde{P}_{A_{2c}} = 1 - \tilde{P}_{A_2}$$

$$= (1 - 0.47, 1 - 0.42, 1 - 0.30, 1 - 0.24)$$

$$= (0.53, 0.58, 0.70, 0.76).$$

Finally, we can write

$$\tilde{P}_T = 1 - \tilde{P}_{X_{1c}} \odot \tilde{P}_{A_{1c}} \odot \tilde{P}_{A_2}$$

$$= (1 - 0.95 \times 0.97 \times 0.76, 1 - 0.93 \times 0.96 \times 0.70,$$

$$1 - 0.88 \times 0.92 \times 0.58, 1 - 0.85 \times 0.90 \times 0.53)$$

$$= (0.30, 0.38, 0.53, 0.59).$$

Thus, we could say that the probability of occurrence of the top event is "between 0.30 and 0.59, with a value most likely to lie between 0.38 and 0.53". Note that if we use exact values for the probabilities of the five fundamental events, we obtain

$$P_{A_1} = P_{X_2} \cdot P_{X_3}$$

$$= 0.20 \times 0.30$$

$$= 0.06$$

$$P_{A_2} = 1 - (1 - P_{X_4})(1 - P_{X_5})$$

$$= 1 - (1 - 0.25)(1 - 0.15)$$

$$= 0.36$$

$$P_T = 1 - (1 - P_{X_1})(1 - P_{A_1})(1 - P_{A_2})$$

$$= 1 - (1 - 0.10)(1 - 0.06)(1 - 0.36)$$

$$= 1 - (0.90)(0.94)(0.64)$$

$$= 1 - 0.54$$

$$= 0.46.$$

It must be observed that the value 0.46 is contained in the prediction "between 0.30 and 0.59" obtained through fuzzy fault-tree analysis. This example demonstrates an important property of fuzzy fault-tree analysis; it is always more predictive than conventional fault tree analysis.

Having obtained the fuzzy probability of the top event, $\tilde{P}_T$, the effectiveness in eliminating each basic event can be computed.

$$\tilde{P}_{T_1} \triangleq \tilde{P}_T \ (0, X_2, X_3, X_4, X_5)$$

$$= 1 - (1 - 0) \odot (1 - \tilde{P}_{A_1}) \odot (1 - \tilde{P}_{A_2})$$

$$= 1 \ (1 - \tilde{P}_{A_1}) \odot (1 - \tilde{P}_{A_2})$$

$$= 1 - \tilde{P}_{A_{1c}} \odot \tilde{P}_{A_{2c}}$$

$$= 1 - 0.97 \times 0.76, \ 1 - 0.96 \times 0.70, \ 1 - 0.92 \times 0.58,$$

$$1 - 0.90 \times 0.53)$$

$$= (0.26, 0.33, 0.47, 0.52).$$

The corresponding improvement index is given by

$$V(\tilde{P}_T, \tilde{P}_{T_1}) = (q_T^\ell - q_{T_1}^\ell) + (p_T^\ell - p_{T_1}^\ell) \ + (p_T^r - p_{T_1}^r)$$

$$+ (q_T^r - q_{T_1}^r)$$

$$= (0.30 - 0.26) + (0.38 - 0.33) + (0.53 - 0.47)$$

$$+ (0.59 - 0.52)$$

$$= 0.22.$$

The effects of eliminating the other basic events can be computed in a similar fashion. Table 3 summarizes the results.

V. APPLICATION OF FUZZY FAULT TREE ANALYSIS

Dust explosions in grain elevators[17] are catastrophic events in the agricultural industry. Whenever grain is handled, a certain amount of breakage invariably occurs and highly combustible grain dust is generated. Oxygen is always present in the atmosphere of a grain storage facility, and a tremendous explosion could result if the mixture of grain dust and air is set off by an ignition source of sufficient energy and duration.

In spite of the number of explosions that have occurred recently and the efforts that have been made to identify their causes, the phenomena involved in a dust explosion are still not well understood. For example, the precise ratio of grain dust and air and the energy and duration of an ignition source that constitute explosive hazards, continue to be the focus of research. Moreover, there is a marked lack of probability data concerning dust explosions. These features render the system a suitable candidate for the application of fuzzy fault tree analysis.

A typical grain storage facility is illustrated in Figure 5. Many dust explosions have been initiated by an explosion or fire in the bucket elevator section (Figure 6). Figure 7 provides the electrical circuit diagram of the head drive system which constitutes a high risk as an ignition source. The fault tree illustrated in Figure 8 has been constructed through an operability study of the bucket elevator section.[18] The top event representing a dust explosion

Table 3
EFFECT OF ELIMINATING BASIC EVENTS

Fuzzy probability	$(q^\ell, p^\ell, p^r, q^r)$	Improvement Index $\tilde{V}(\tilde{P}_T, \tilde{P}_{Ti})$
$\tilde{P}_T$	(0.30, 0.38, 0.53, 0.59)	—
$\tilde{P}_{T_1}$	(0.26, 0.33, 0.47, 0.52)	0.22
$\tilde{P}_{T_2}$	(0.28, 0.35, 0.49, 0.55)	0.13
$\tilde{P}_{T_3}$	(0.28, 0.35, 0.49, 0.55)	0.13
$\tilde{P}_{T_4}$	(0.12, 0.20, 0.35, 0.43)	0.70
$\tilde{P}_{T_5}$	(0.26, 0.30, 0.41, 0.46)	0.37

or fire in the bucket elevator has three necessary conditions: (1) the presence of air, (2) the presence of ignitable dust, and (3) the presence of an ignition source. In this work, we focus only on the existence of an ignition source; this corresponds to the top event T_1. The fundamental and intermediate events leading to T_1 are presented in Tables 4 and 5, respectively.

The present study utilizes rough estimates of the fundamental event probabilities supplied by operations personnel. These estimates are conveniently expressed in the form of trapezoidal fuzzy sets (Table 6). The fuzzy probabilities are propagated up the fault tree by employing the approximate product and complement operators. The resultant fuzzy probability of occurrence of an ignition source is given by

$$\tilde{P}_{T_1} = (0.051, 0.105, 0.188, 0.249).$$

This corresponds to a probability most likely to fall in the range between 0.105 and 0.188; however, it can be as low as 0.051, and is no greater than 0.249.

The minimal cut sets for the fault tree together with the corresponding probability of the top event are presented in Table 7. The effectiveness of eliminating basic events can be gauged by computing the index of improvement. The results given in Table 8 demonstrate that substantial reductions in top event probability can be achieved by eliminating events X_6 (operator failure in stopping the head drive motor) and X_1 (welding, cutting, or fire from other sections). Both these events are related to human errors. Their occurrence can be prevented by good management and proper training of personnel.

VI. CONCLUSIONS

One of the major problems facing many quantitative risk assessment studies is the lack of proper data about the probabilities of occurrence of hazardous events. Often, the only alternative is to employ rough estimates provided by design and operations personnel. Although the estimates are usually expressed in imprecise linguistic terms, they represent an important part of human expertise in the domain. It is worthwhile to attempt to represent and engage this knowledge.

Fuzzy sets provide a convenient approach for coping with the task of expressing imprecise or approximate notions of probability. Trapezoidal fuzzy sets are reasonably effective for modeling approximate concepts; more importantly, they facilitate the computation of fuzzy probabilities. Also, the use of the approximate product operator instead of the true product defined by the extension principle is not a severe disadvantage. The approximation tends to provide a margin of safety, which in any case, is rather small.

Perhaps the most important advantage of a fuzzy fault tree approach is its ability to engage

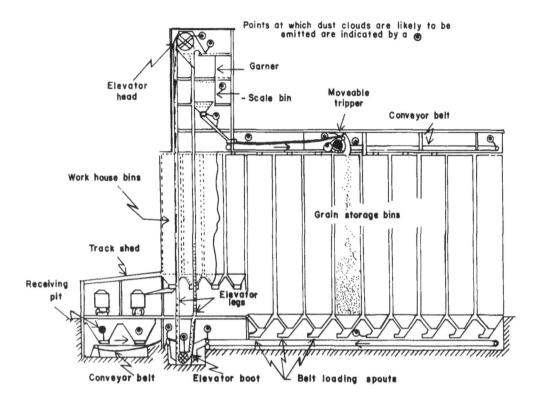

FIGURE 5. Schematic diagrams of a typical grain elevator.

human expertise in an assessment of risk. Human beings often possess quality knowledge that arises out of their insights and experiences. When the safety of a system is being examined this expertise has the greatest weight.

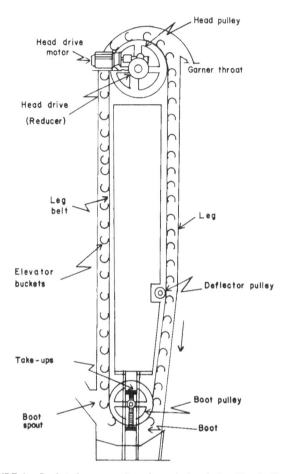

FIGURE 6.   Bucket elevator section of a typical grain handling facility.

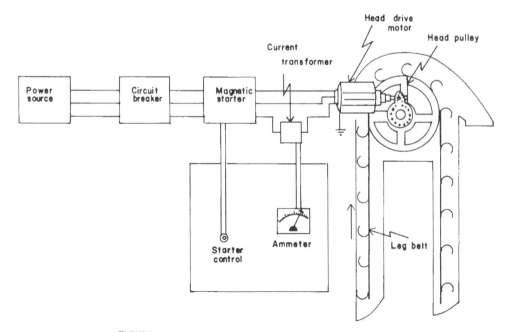

FIGURE 7.   Electrical circuit diagram of head drive system.

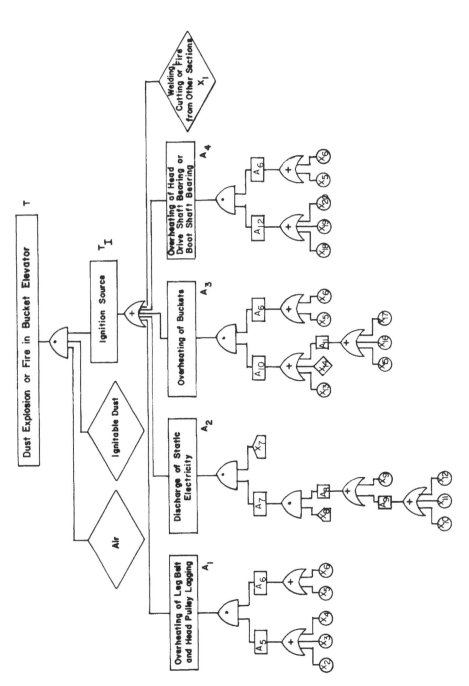

FIGURE 8.   Fault tree for a dust explosion or fire in the bucket elevator.

Table 4
FUNDAMENTAL EVENTS LEADING TO AN IGNITION SOURCE

X_1:	Welding, cutting, or fire from other sectors
X_2:	Maladjustment of the boot shaft take-ups
X_3:	Wear and tear of the head pulley lagging
X_4:	Overloading of the leg
X_5:	Failure of the head drive motor to stop due to failure of the magnetic starter
X_6:	Failure of the operator to stop the head drive motor
X_7:	Presence of spark gap
X_8:	Generation of static electricity in the leg
X_9:	Low humidity in the leg
X_{10}:	Use of nonconductive lubricant
X_{11}:	Breaking of the grounding wire
X_{12}:	Improper installation of the grounding wire
X_{13}:	Malfunction of the buckets
X_{14}:	Foreign material in the leg
X_{15}:	Malfunction of the leg belt
X_{16}:	Malfunction of the boot shaft take-ups
X_{17}:	Maladjustment of the boot shaft take-ups
X_{18}:	Inadequate lubrication
X_{19}:	Failure of the bearings
X_{20}:	Overload of the bearings

Table 5
INTERMEDIATE EVENTS

A_1:	Overheating of the leg belt and heat pulley lagging
A_2:	Discharge of static electricity
A_3:	Overheating of the buckets
A_4:	Overheating of the head drive shaft bearings or boot shaft bearings
A_5:	Slippage between the leg belt and head pulley
A_6:	Continuing operation of the head drive motor
A_7:	Accumulation of static electricity in the leg
A_8:	Inadequate leakage of static electricity from the leg
A_9:	Poor grounding of the head shaft, boot shaft, or deflector shaft
A_{10}:	Friction between the inside walls of the leg and buckets
A_{11}:	Low tension in the leg belt
A_{12}:	Malfunction of the head drive shaft bearings or boot shaft bearings

Table 6
FUZZY PROBABILITIES OF
OCCURRENCE OF THE
FUNDAMENTAL EVENTS

Event X_i	Fuzzy probabilities $\tilde{P}_{Xi}$
X_1	(0.010, 0.030, 0.060, 0.080)
X_2	(0.050, 0.060, 0.080, 0.100)
X_3	(0.080, 0.100, 0.120, 0.140)
X_4	(0.100, 0.120, 0.180, 0.200)
X_5	(0.010, 0.020, 0.030, 0.040)
X_6	(0.050, 0.070, 0.090, 0.100)
X_7	(0.010, 0.020, 0.040, 0.050)
X_8	(0.200, 0.250, 0.350, 0.400)
X_9	(0.100, 0.150, 0.250, 0.300)
X_{10}	(0.005, 0.007, 0.008, 0.010)
X_{11}	(0.100, 0.120, 0.130, 0.140)
X_{12}	(0.005, 0.007, 0.009, 0.010)
X_{13}	(0.010, 0.020, 0.030, 0.040)
X_{14}	(0.150, 0.200, 0.300, 0.350)
X_{15}	(0.050, 0.060, 0.070, 0.080)
X_{16}	(0.010, 0.020, 0.030, 0.040)
X_{17}	(0.010, 0.020, 0.030, 0.040)
X_{18}	(0.100, 0.120, 0.140, 0.160)
X_{19}	(0.050, 0.060, 0.070, 0.080)
X_{20}	(0.080, 0.100, 0.120, 0.140)

Table 7
MINIMAL CUT SETS AND THE
CORRESPONDING FUZZY PROBABILITY OF
THE TOP EVENT

Minimal cut set	Fuzzy probability of the top event × 1000	
X_2, X_5	(0.500, 1.200, 2.400, 4.000)	
X_2, X_6	(0.500, 4.200, 7.200, 10.000)	
X_3, X_5	(0.800, 2.000, 3.600, 5.600)	
X_3X_6	(4.000, 7.000, 10.800, 14.000)	A_1
x_4, x_5	(1.000, 2.400, 5.400, 8.000)	
X_4, X_6	(5.000, 8.400, 16.200, 20.000)	
X_7, X_8, X_9	(0.200, 0.750, 3.500, 6.000)	
X_7, X_8, X_{10}	(0.010, 0.035, 0.112, 0.200)	
X_7, X_8, X_{11}	(0.200, 0.600, 1.820, 2.800)	A_2
X_7, X_8, X_{12}	(0.010, 0.035, 0.126, 0.200)	
X_5, X_{13}	(0.100, 0.400, 0.900, 1.600)	
X_5, X_{14}	(1.500, 4.000, 9.000, 14.000)	
X_5, X_{15}	(0.500, 1.200, 2.100, 3.200)	
X_5, X_{16}	(0.100, 0.400, 0.900, 1.600)	
X_5, X_{17}	(0.100, 0.400, 0.900, 1.600)	
X_6, X_{13}	(0.500, 1.400, 2.700, 4.000)	A_3
X_6, X_{14}	(7.500, 14.000, 27.000, 35.000)	
X_6, X_{15}	(2.500, 4.200, 6.300, 8.000)	
X_6, X_{16}	(0.500, 1.400, 2.700, 4.000)	
X_6, X_{17}	(0.500, 1.400, 2.700, 4.000)	

Table 7 (continued)
MINIMAL CUT SETS AND THE CORRESPONDING FUZZY PROBABILITY OF THE TOP EVENT

Minimal cut set	Fuzzy probability of the top event × 1000	
X_5, X_{18}	(1.000, 2.400, 4.200, 6.400)	
X_5, X_{19}	(0.500, 1.200, 2.100, 3.200)	
X_5, X_{20}	(0.800, 2.000, 3.600, 5.600)	
X_6, X_{18}	(5.000, 8.400, 12.600, 16.000)	A_4
X_6, X_{19}	(2.500, 4.200, 6.300, 8.000)	
X_6, X_{20}	(4.000, 7.000, 10.800, 14.000)	
X_1	(10.000, 30.000, 60.000, 80.000)	

Table 8
FUZZY PROBABILITIES AND IMPROVEMENT INDEXES

$\tilde{P}$	Fuzzy probability × 1000 $(q^\ell, p^\ell, p^r, q^r)$	Improvement index × 1000 $\tilde{V}(\tilde{P}_{T_I}, \tilde{P}_{T_{li}})$	Rank
$\tilde{P}_{TI}$	(51.0, 105.0, 188.0, 249.0)	—	—
$\tilde{P}_{TI1}$	(40.9, 77.8, 136.6, 183.3)	154.3	2
$\tilde{P}_{TI2}$	(47.7, 100.6, 180.6, 238.0)	26.0	9
$\tilde{P}_{TI3}$	(46.0, 97.3, 176.6, 233.7)	39.3	8
$\tilde{P}_{TI4}$	(44.8, 95.6, 170.5, 227.1)	54.9	5
$\tilde{P}_{TI5}$	(44.1, 89.5, 159.3, 206.2)	93.8	3
$\tilde{P}_{TI6}$	(17.2, 48.3, 97.6, 137.3)	292.5	1
$\tilde{P}_{TI7}$	(50.1, 104.2, 183.8, 241.7)	13.1	12
$\tilde{P}_{TI8}$	(50.1, 104.2, 183.8, 241.7)	13.1	12
$\tilde{P}_{TI9}$	(50.3, 104.8, 185.5, 244.1)	8.2	17
$\tilde{P}_{TI10}$	(50.5, 105.4, 188.2, 248.5)	0.3	19
$\tilde{P}_{TI11}$	(50.3, 104.8, 186.8, 246.5)	4.5	18
$\tilde{P}_{TI12}$	(50.5, 105.4, 188.2, 248.5)	0.3	19
$\tilde{P}_{TI13}$	(49.9, 103.9, 185.5, 244.4)	9.2	14
$\tilde{P}_{TI14}$	(41.9, 89.1, 158.4, 210.4)	93.1	4
$\tilde{P}_{TI15}$	(47.7, 100.5, 181.5, 240.2)	23.0	10
$\tilde{P}_{TI16}$	(49.9, 103.9, 185.5, 244.4)	9.2	14
$\tilde{P}_{TI17}$	(49.9, 103.9, 185.5, 244.4)	9.2	14
$\tilde{P}_{TI18}$	(44.8, 95.6, 174.5, 231.5)	46.5	6
$\tilde{P}_{TI19}$	(47.7, 100.6, 181.5, 240.1)	23.0	10
$\tilde{P}_{TI20}$	(46.0, 97.3, 176.6, 233.6)	39.4	7

REFERENCES

1. **Haasl, D. F.**, Advanced concepts in fault tree analysis, *Proceedings of the System Safety Symposium*, The Boeing Company, Seattle, 1965.
2. **Barlow, R. E. and Chatterjee, P.**, Introduction to Fault Tree Analysis, Rep. OCR 73-70, Operations Research Center, University of California, Berkeley, 1973.
3. **Barlow, R. E. and Lambert, H. E.**, Introduction to fault tree analysis, in *Reliability and Fault Tree Analysis — Theoretical and Applied Aspects of System Reliability and Safety Assessment*, Barlow, R. E., Fussell, J. B., and Singpurwalla, N. D., Society for Industrial and Applied Mathematics, Philadelphia, 1975, 7.
4. **Fussell, J. B.**, Fault tree analysis: concepts and techniques, in *Generic Techniques in Systems Reliability Assessment*, Henley, E. J. and Lynn, J. W., Eds., Noordhoff, Leyden, 1976, 133.
5. **Henley, E. J. and Kumamoto, H.**, *Designing for Reliability and Safety Control*, Prentice-Hall, Englewood Cliffs, N.J., 1985.
6. **Zadeh, L. A.**, Fuzzy sets, *Inform. Control.*, 8, 338, 1965.
7. **Kaufmann, A.**, *Introduction to the Theory of Fuzzy Subsets*, Vol. 1, Academic Press, New York, 1975.
8. **DuBois, D. and Prade, H.**, *Fuzzy Sets and Systems: Theory and Applications*, Academic Press, New York, 1980.
9. **Noma, K., Tanaka, H., Asai, K.**, On fault tree analysis with fuzzy probability, *J. Ergon.*, 17, 291, 1981.
10. **Tanaka, H., Fan, L. T., Lai, F. S., and Toguchi, K.**, Fault tree analysis by fuzzy probability, *IEEE Trans. Reliability*, R-32(5), 453, 1983.
11. **Lawley, H. G.**, Operability study and hazard analysis, *Chem. Eng. Progr.*, 70(4), 58, 1974.
12. Chemical Industry Safety and Council of the Chemical Industries Association, *A Guide to Hazard and Operability Studies*, Alembic House, London, 1977.
13. **Knowlton, R. E.**, *An Introduction to Hazard and Operability Studies*, Chemetics International, Vancouver, 1981.
14. **Kletz, T. A.**, Eliminating potential process hazards, *Chem. Eng.*, 97(7), 48, 1985.
15. **Fussell, J. B. and Velsely, W. E.**, A new method for obtaining cut sets, *Trans. Am. Nucl. Soc.*, 15(1), 262, 1972.
16. **Zadeh, L. A.**, The concept of a linguistic variable and its application to approximate reasoning. I, II, III., *Inform. Sci.*, 8, 199; 8, 301; 9, 43, 1975.
17. **Kameyama, Y., Lai, F. S., Sayama, H., and Fan, L. T.**, The risk of dust explosions in grain processing and handling facilities, *J. Agric. Eng. Res.*, 27, 253, 1982.
18. **Lai, F. S., Aldis, D. F., Kameyama, Y., Sayama, H., and Fan, L. T.**, Operability study of a grain processing and handling facility, *Trans. ASAE*, 2(1), 222, 1984.

INDEX

A

Abnormal motion command, 74
Abnormal trajectory, 74
Acceptable risks, 13
Accomplice, 66
Actuator failure, 72
Acute risks, 2, 4—7
Air-cut, 98, 101
Air-run, 100—101
Air-to-fuel, 100
"AND" gate, 24, 29, 37—38, 41, 119—121
AND/OR tree, 83—84, 87—89, 94, 110—114, 118
Application software, 83
Artificial intelligence, 83, 84
Asbestos, 5
Asbestosis, 5—6
"As good as new" components, 33
Assailant, 66
Associated binary function, 39
Asymptotic unavailability, 32
Attitude controllers, 72, 75
Authorized intrusion, 74
Automatic motion interlock, 78—79
Availability, 30
Average unavailability, 33

B

Background factors, 6, 67, 69
Backward chaining, 113
Barrier LTA, 72, 74
Barriers, 29, 72, 74, 75
BASIC, 85
Basic events, 22, 24, 35, 37—38, 64, 78—79, 129
Bathtub curve, 32
Benzene, 6
β-factor method, 34
Binary functions, 37—39
Binomial distribution, 32
Bottom-up approach, 40
Breadth first search, 112
Byssinosis, 6

C

Cancer, 6—7
CASE, 85
Causal models, 64—65, 70—75
Cause consequence diagrams, 83
Character analysis, 62, 66—67
Checklists, 83, 88, 119—120
Chemical hazards, 5—6, 25
Child-state-name, 110
Chronic risks, 2, 4, 7
Coal dust, 3—4

Codd's theory, 83
Command failure, 93
Common external cause, 34
Common mode data, 33—34
Comparison of risks, 1—9
Complete air-run, 99, 102
Component failure modes, 26, 30—33, 69
Comprehensive causal models, 70—75
Confirmation, 111, 114
Consistency maintenance, 97—98
Constant failure probabilities, 30
Continuous motion, 77
Controlled motions, 68
Control limits, 2, 5, 13
Correct information blockage, 67
Corrective maintenance, 32
Corrective measures in robot systems, 65
Costs, 12—18
Cut set, 94, 122

D

Dangerous zone, authorized and unauthorized intrusion, 74, 77
  human in, see Human in dangerous zone
Database, 82—85, 114
Data manipulation language (DML), 83—85, 110, 115
Data-oriented inferences, 112, 114
Data-oriented search, 85
dBASE, 83, 85
dBASE III, 84
Death, risk of, 4
Debug, 114
Declarative languages, 85
Deductive approach, 63—65
Defective component, 88
Defense mechanism, 6
Delayed death, 2, 4, 7
DENDRAL, 82
Direct stimulation, 39—40, 44
Dispersion, 37
DML, see Data manipulation language
DNA database, 84
Domain expert, 82
Domestic activities, cost of saving a life, 16
Domino effect, 66
Dose-response relationship, 3—6
Double-correlation scenario, 66
Downtime, 32
Dual tree, 120—122
Dust explosions in grain elevators, 128—136

E

Energy blockage, 66, 68

F

G

H

Printed and bound by CPI Group (UK) Ltd, Croydon, CR0 4YY

22/10/2024

01777632-0011